Functional Nanomaterials for Biosensors and Biomedicine Application

Functional Nanomaterials for Biosensors and Biomedicine Application

Guest Editors

Liqiang Luo
Xiaoqiang Liu

MDPI

Basel • Beijing • Wuhan • Barcelona • Belgrade • Novi Sad • Cluj • Manchester

Guest Editors

Liqiang Luo
Department of Chemistry
Shanghai University
Shanghai
China

Xiaoqiang Liu
College of Chemistry and
Molecular Sciences
Henan University
Kaifeng
China

Editorial Office
MDPI AG
Grosspeteranlage 5
4052 Basel, Switzerland

This is a reprint of the Special Issue, published open access by the journal *Molecules* (ISSN 1420-3049), freely accessible at: https://www.mdpi.com/journal/molecules/special_issues/290C412IEX.

For citation purposes, cite each article independently as indicated on the article page online and as indicated below:

Lastname, A.A.; Lastname, B.B. Article Title. *Journal Name* **Year**, *Volume Number*, Page Range.

ISBN 978-3-7258-7587-0 (Hbk)
ISBN 978-3-7258-7588-7 (PDF)
https://doi.org/10.3390/books978-3-7258-7588-7

Contents

About the Editors

Liqiang Luo

Liqiang Luo is a professor in the Department of Chemistry, College of Sciences, Shanghai University. He received his BS and MS degrees from Henan Normal University, China, in 1994 and 1997, respectively, and his PhD from the Changchun Institute of Applied Chemistry, Chinese Academy of Sciences, China, in 2000. After that, he worked as a postdoctoral researcher at the Nanoarchitectonics Research Center, National Institute of Advanced Industrial Science and Technology (AIST), Tsukuba, Japan. His current research is focused on the synthesis, characterization and application of new functional nanomaterials. He has published more than 200 research papers and reviews.

Xiaoqiang Liu

Xiaoqiang Liu is a professor at Henan University, and Distinguished Professor of the Yellow River Scholar Program. He received his PhD degree from Macquarie University, Australia, in 2009. His research interests include the preparation of novel functional materials for the construction of sensors and energy devices. He has presided over projects including the General Program of the National Natural Science Foundation of China, the Joint Fund of the National Natural Science Foundation of China, the Ministry of Education Scholarship Fund, the Program for Science and Technology Innovation Talents in Universities of Henan Province, and the Henan International Science and Technology Cooperation Cultivation Project. His research mainly focuses on the applications of novel functional materials in electrocatalysis, photoelectrochemical catalysis, sensors and energy-related fields. As the first author and corresponding author, he has published more than 50 academic papers in top-tier SCI journals, such as *Applied Catalysis B: Environment and Energy*, *Analytical Chemistry*, *Chemical Communications*, *the Journal of Materials Chemistry A*, *ACS Applied Materials & Interfaces*, *Biosensors & Bioelectronics*, *Chemical Engineering Journal*, and *Small*.

molecules

MDPI

cle

exible Copper-Based TEM Grid for Microscopic haracterization of Aged Magnetotactic Bacteria MS-1 and ıeir Magnetosome Crystals in Air-Dried Droplet

:alia Lorela Paul [1,2,3], Regis Deturche [2,3], Jeremie Beal [2,3], Catalin Ovidiu Popa [1,3] and lica Elena Ionescu [2,3,*]

1 Materials Science and Engineering Department, Faculty of Materials and Environmental Engineering, Technical University of Cluj-Napoca, 400641 Cluj-Napoca, Romania

2 Light, Nanomaterials and Nanotechnologies (L2n) Laboratory, CNRS UMR 7076, University of Technology of Troyes, 12 Rue Marie Curie, CS 42060, CEDEX, 10004 Troyes, France

3 Eut+ Institute for Nanomaterials & Nanotechnologies EUTINN, European University of Technology, European Union

* Correspondence: elena_rodica.ionescu@utt.fr

Abstract

Magnetotactic bacteria (MTB) have attracted interest in recent years, mainly due to their natural ability to form intracellular magnetic nanocrystals with potential for biomedical and environmental applications. In this study, we focused on the morphological analysis of the *Paramagnetospirillum magnetotacticum* MS-1 strain, trying to keep the bacteria as close to their natural state as possible. An important element of this work is the use of untreated bacterial cells, without conductive coating or chemical fixation, using a simple and low-cost support. This choice was made intentionally to avoid changes induced by metallization and to allow direct observation of characteristics that may be relevant in applications where the interaction of the bacteria with the environment plays an important role, such as biosensors. In addition, the analysis was performed on a bacterial suspension stored for approximately 10 months at 4 °C to assess whether the morphology specific to the MS-1 strain is maintained over time. The obtained results show that the general cell morphology and magnetosome organization can be clearly and reproducibly observed even after long-term storage. Without attempting to replace studies based on conventional sample preparation methods, this work provides a complementary perspective and suggests that magnetotactic bacteria may represent a natural and effective alternative to synthetic magnetic nanoparticles, with potential applications in the biomedical and environmental fields.

Keywords: magnetotactic bacteria; *Paramagnetospirillum magnetotacticum* MS-1; magnetosomes; natural magnetic nanoparticles; drop-casting method; flexible TEM grid; microscopic techniques

1. Introduction

Magnetotactic bacteria represent a diverse group of prokaryotic, Gram-negative, and motile microorganisms found mainly in aquatic environments, particularly at the oxic–anoxic interface in water columns or sediments. These bacteria move using flagella and have the ability to orient themselves along the lines of the Earth's magnetic field, a behavior known as magnetotaxis. This property gives them an important adaptive advantage, allowing them to more efficiently locate areas with optimal oxygen conditions in chemically and redox-stratified environments [1–4].

A defining feature of MTB is the presence of magnetosomes, biomineralized structures formed from magnetic nanocrystals of magnetite (Fe_3O_4) or greigite (Fe_3S_4), enclosed in a phospholipid membrane. The relatively uniform dimensions of these crystals, typically between 35 and 120 nm, give them magnetic stability and a permanent magnetic moment, making them single-domain particles. Magnetosomes are usually organized in intracellular chains, which allows bacteria to passively align with magnetic field, similar to a small magnetic needle [5–9]. Due to their high crystalline quality, natural biocompatibility, and magnetic stability, magnetosomes have significant advantages over synthetic magnetic nanoparticles [5,8,10].

In the context of increased interest in biological materials usable in nanotechnology, biomedicine, and environmental applications, MTBs have been proposed as natural platforms for biosensors, medical imaging, or controlled drug delivery systems [5,11,12]. However, the exploitation of these applications depends directly on a correct structural characterization of bacteria and magnetosomes. Conventional methods of sample preparation for microscopy often involve fixation, dehydration, or metallization, procedures that can alter cell morphology and surface properties, removing samples from their native state and introducing unwanted artifacts.

Paramagnetospirillum magnetotacticum MS-1 is a Gram-negative, microaerophilic, spiral-shaped magnetotactic bacterium that synthesizes magnetite with dimensions of approximately 45 nm [13–15]. The aim of this study was to characterize the morphology and structure of the MS-1 strain using a simple and accessible approach based on the analysis of untreated cells deposited on a low-cost copper-based TEM grid. Through this strategy, the paper aims to evaluate the feasibility of a rapid microscopic analysis method, designed to reflect as accurately as possible the natural state of the bacteria and to provide complementary information to studies based on classical sample preparation methods.

2. Results

2.1. SEM and STEM Analyses of Copper TEM Grid

2.1.1. Untreated with MS-1 Bacterial Cells

Detailed analysis of the morphology, crystal structure, chemical composition, and organization of magnetosomes requires the use of advanced microscopic techniques that allow visualization of the bacterial surface and the distribution of magnetosomes, as well as information regarding the mapping of chemical elements or aspects related to the topography of the bacterial surface. In this regard, we analyzed the bacteria using SEM and STEM, with the aim of studying the external cellular morphology, identifying the shape of the bacteria and their size but also attempting to visualize the magnetosome chain visible on the cell surface.

In this analysis, a single TEM-grid support was used through several analysis techniques (SEM, STEM, EDX, and later AFM) to understand its characteristics and composition as a feasible support in MTB analysis.

SEM and STEM investigations, as shown in Figure 1, illustrate the morphology of a copper-based TEM grid used as a sample support, before the deposition of MS-1 bacterial cells. The SEM images highlight the surface, free of contaminants or residual particles, confirming that the grid is untreated and in the absence of artifacts that could interfere with subsequent bacterial sample analysis.

https://doi.org/10.3390/molecules310

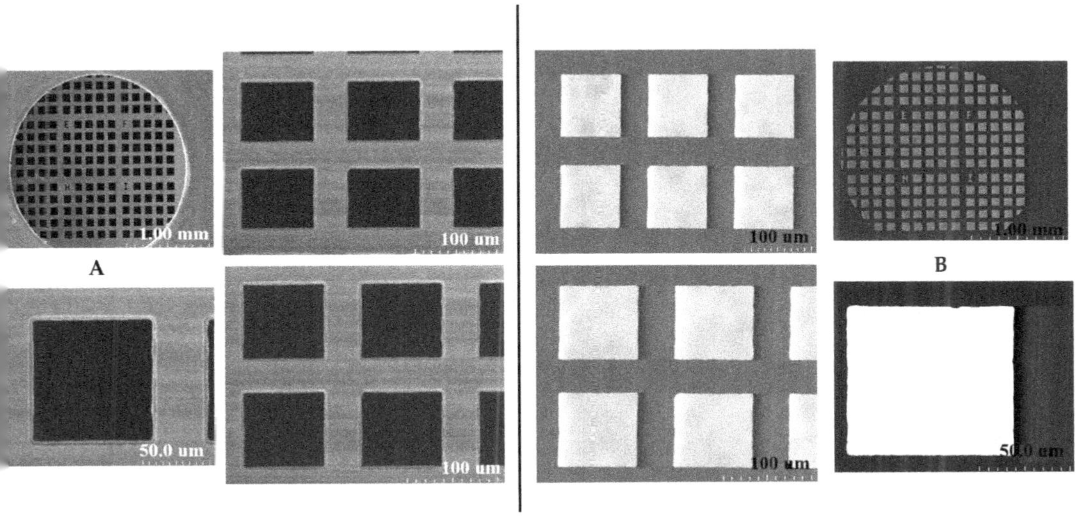

Figure 1. SEM (**A**) and STEM (**B**) imaging of an untreated copper-based TEM grid.

2.1.2. Treated with MS-1 Bacterial Cells

The analysis of MS-1 by SEM allowed the detailed characterization of the morphology of MTB and highlighted variations in their shape and size. The MS-1 bacteria have a small degree of morphological diversity; they have mainly spiral forms, but some of them are more elongated (Figure 2): morphology (1) more elongated, compared to morphology (2) with typical spiral shape. The cell surface appears relatively smooth, and at some of them it was possible to identify flagella. The dimensions of the cells analyzed by SEM vary; the length oscillates from a few hundred nanometers up to a few micrometers. As can be seen in Figure 2, the size of the bacteria varies between 1.62 and 5.39 μm, while their width varies between 250 nm and 3.00 μm. The ratio between length and width indicate the differences between elongated and more spiral-shaped bacteria, highlighting the morphological plasticity of these microorganisms.

Magnetosomes visualized by SEM also show variations in size and number in the constitution of the chains. Magnetosomal particles generally fall within the range of 30–120 nm [16–18], considered optimal for magnetotaxis, and the shape is predominantly cuboctahedral. The present measurements fell within the range between 30 and 76 nm for the MS-1 strain of MTB.

The distribution of magnetosome particles in chains within the same bacterial population indicates a considerable dispersion, both in chain length and in organization and distribution within the cell (13–34 magnetosome crystals per cell). This aspect can be explained by adaptation to environmental factors, such as oxygen concentration or nutritional components in the environment, which have direct implications on the organization of magnetosome chains, on magnetic coercion, and the cellular magnetic moment.

In this way, the characterization of these bacteria by SEM not only highlights the morphological and structural diversity of MTB, but also provides essential data for understanding the relationship between shape, size, and functionality within this group of microorganisms.

https://doi.org/10.3390/molecules31020253

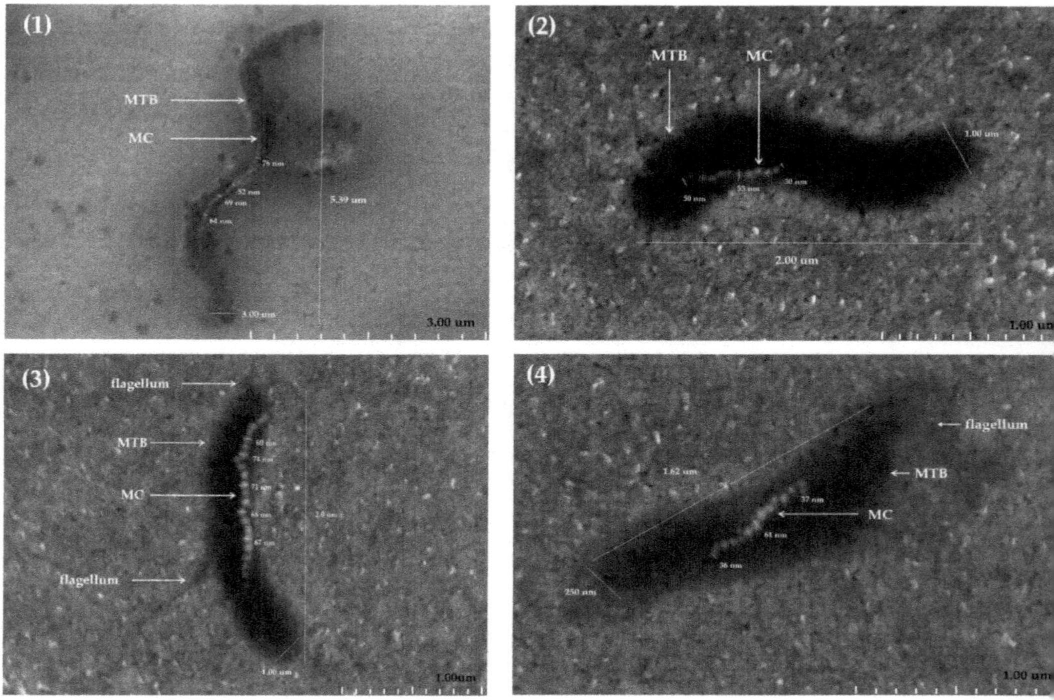

Figure 2. SEM morphology of four (**1–4**) MS-1 bacteria (sizes between 1.62–5.39 µm) and their magnetosome chains (13–34 magnetosome crystals per cell with sizes of crystals between 36 and 76 nm); MTB—magnetotactic bacteria, MS-1 strain; MC—magnetosome chain.

In STEM, images are obtained by focusing a very thin beam of electrons, which is systematically scanned along an extremely thin sample. The beam interacts with the sample, and the resulting electrons are collected by detectors, allowing an image to be formed, and can reflect both the morphological structure and the internal distribution of the material [19].

In the present context, the copper-based TEM grid was used to assess the feasibility of this type of support in MTB, MS-1 strain analysis. The substances dissolved in the culture medium formed a compact matrix in the holes of the TEM grid, allowing STEM analysis. As can be seen in Figure 3A, several bacteria could be identified, their distribution and most importantly the dispersed way in which they are located can also be observed. The essential aspect that can be deduced is that the bacteria are found rather isolated than in clusters. The bacteria analyzed in STEM (Figure 3B) presents an elongated morphology, with a length of 3.94 µm. STEM analysis allows clear visualization of the magnetosome chain, the organized distribution inside the cell, as well as the appreciation of its length (28 magnetic crystals). The analyzed nanocrystals have dimensions between 30 and 50 nm. Thus, STEM analysis of MS-1 revealed elongated cells with a helical morphology characteristic of the genus. The images obtained allowed the clear observation of magnetosome chains arranged orderly along the longitudinal axis of the cell. These structures appeared as nanoparticles with uniform dimensions and high contrast (Figure 3).

https://doi.org/10.3390/molecules31

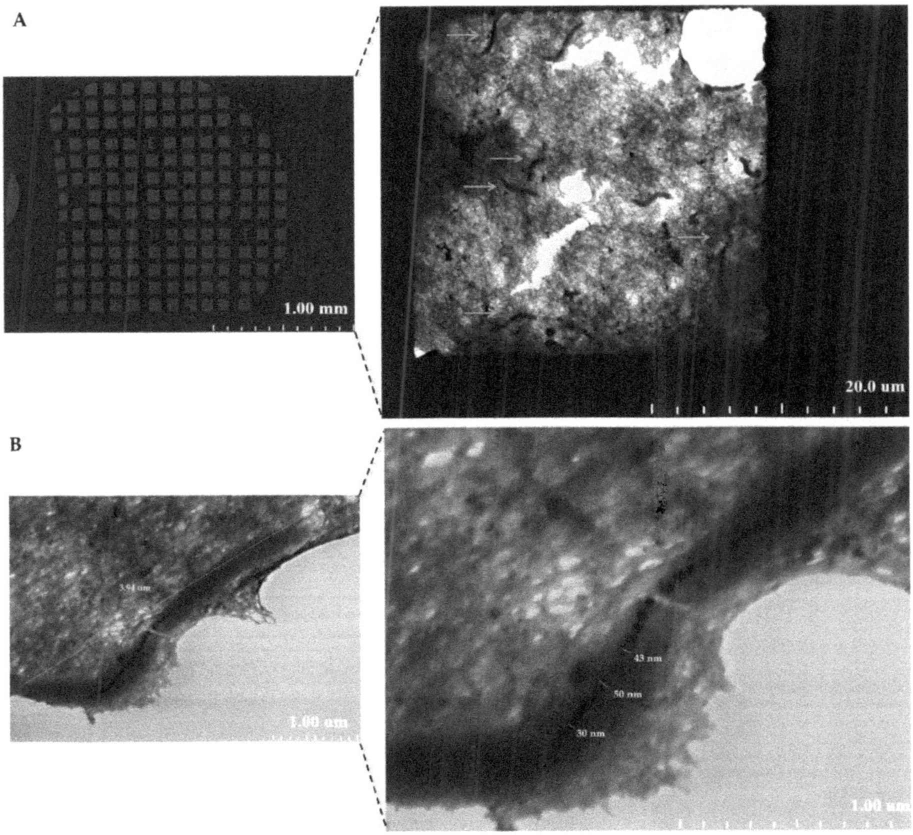

Figure 3. STEM image of MS-1 bacteria; (**A**)—STEM imaging of TEM grid with bacteria distributed on it, (**B**)—STEM imaging of MS-1 (length of 3.94 µm) and their magnetosome chain (28 magnetic crystals with sizes between 30 and 50 nm); bacteria indicated by orange arrows, MC—magnetosome chain.

2.2. *EDX Analysis of a Copper TEM Grid*

2.2.1. Untreated with MS-1 Bacterial Cells

EDX analysis performed on the untreated copper grid, as shown in Figure 4, revealed a spectrum dominated by the characteristic signal of copper, confirming the basic composition of the support used. In the spectrum, moderate intensity signals for sodium and weaker signals for iron, calcium, and chlorine, as well as for carbon, oxygen, aluminium, and magnesium were also observed. The presence of carbon and oxygen can be explained by the existence of the thin carbon film of the grid and by the adsorption of compounds from the ambient air, while sodium, calcium, and chlorine suggest possible saline residues from the handling of the grid. Iron, aluminium, and magnesium can be attributed to minor contamination with metal particles or dust, occurring during the manufacturing process or handling. The low intensities of these signals indicate that they represent a minimal background, which does not compromise the subsequent interpretation of the data [20–22]. Establishing this elemental profile is important because it provides a reference standard for the analysis of biological samples subsequently loaded onto the grid, allowing for the clear distinction of contributions from bacterial cells and magnetosomes from possible surface contaminations.

https://doi.org/10.3390/molecules31020253

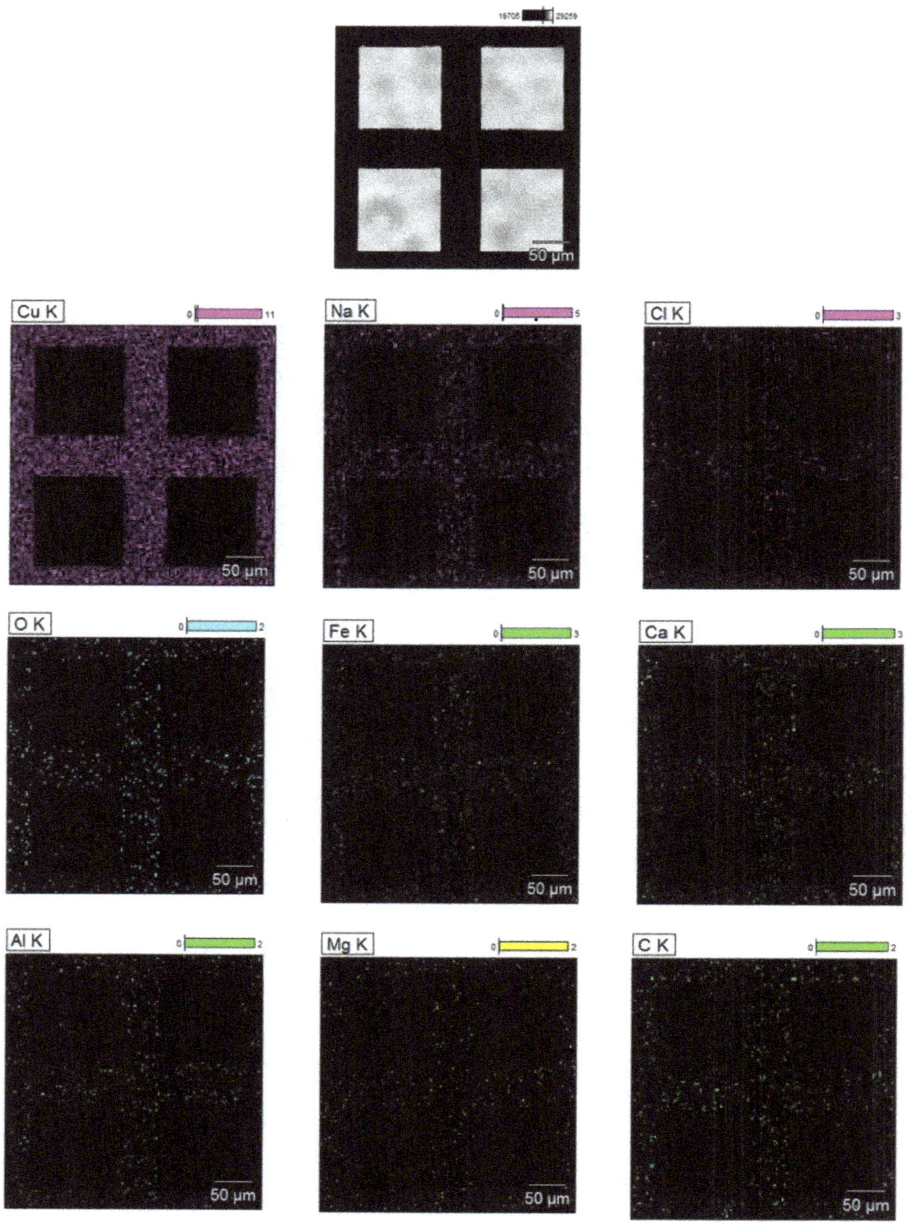

Figure 4. SEM image and EDX maps obtained on an untreated copper TEM grid; the square areas that appear black correspond to the gaps (mesh) in the grid, where no material is present and no EDX signal is detected. The colored areas in the EDX maps represent the solid edges of the grid, where the characteristic elements of the support are detected.

2.2.2. Treated with MS-1 Bacterial Cells

EDX analysis of MS-1 allowed the determination of the elemental distribution of intracellular components, in particular magnetosomes, which contain iron-rich nanocrystals. By detecting the X-rays emitted by the sample, following electron beam excitation, EDX

https://doi.org/10.3390/molecules310

provided precise information on the chemical composition of intracellular structures, as shown in Figure 5. In the case of MS-1 magnetosomes, the spectrum obtained by X-ray point scan highlights the presence of iron and oxygen, confirming the controlled biomineralization of magnetite.

A

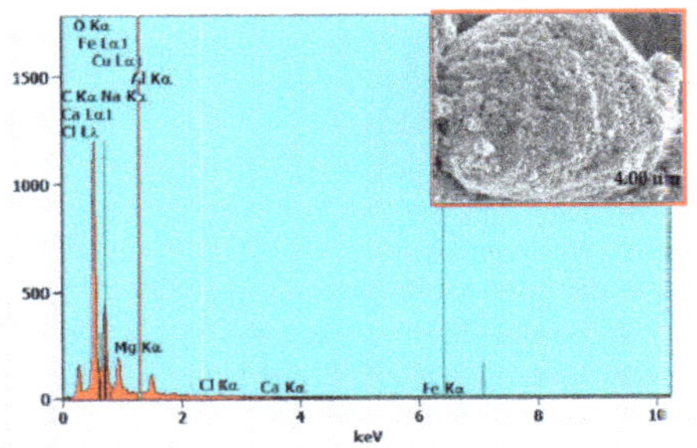

Element Line	Weight %	Atom %
C K	2.3	5.6
O K	31.3	56.8
Na K	0.0	0.0
Mg K	0.5	0.6
Al K	4.9	5.3
Cl K	0.0	0.0
Ca L	0.0	0.0
Fe L	61.0	31.7
Total	100.0	100.0

B

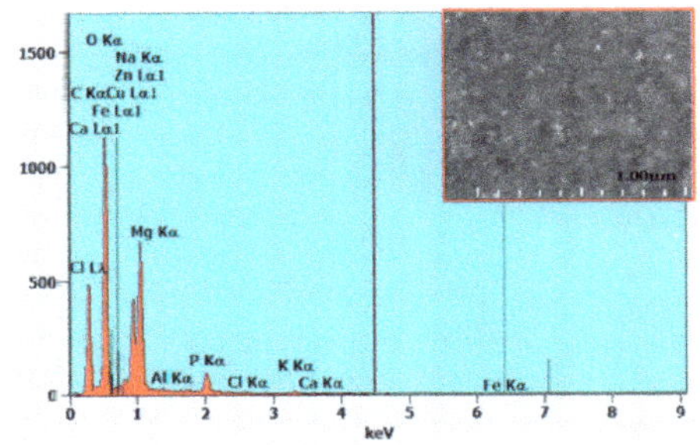

Element Line	Weight %	Atom %
C K	8.2	16.7
O K	28.2	43.0
Na K	18.8	20.0
Mg K	0.0	0.0
Al K	0.0	0.0
P K	6.7	5.3
Cl K	0.6	0.4
Ca L	0.0	0.0
Fe L	4.1	1.8
Cu L	24.3	9.3
Zn L	9.0	3.4
Total	100.0	100.0

C

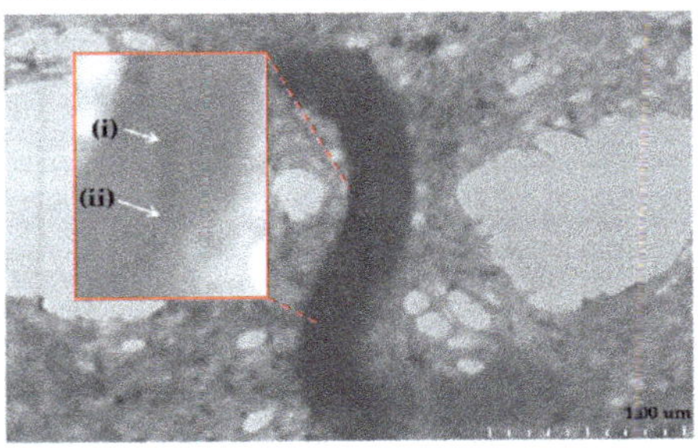

Element Line	Weight %		Atom %	
	(i)	(ii)	(i)	(ii)
C K	20.6	22.6	34.2	39.5
N K	7.8	3.1	11.1	4.7
O K	28.6	28.1	35.7	36.8
Na K	6.1	2.5	5.3	2.2
Al K	1.9	1.7	1.4	1.3
P K	1.2	1.4	0.8	0.9
K K	0.9	0.6	0.5	0.3
Fe K	17.9	24.2	6.4	9.1
Cu K	14.8	15.8	4.6	5.2
Total	100%	100%	100%	100%

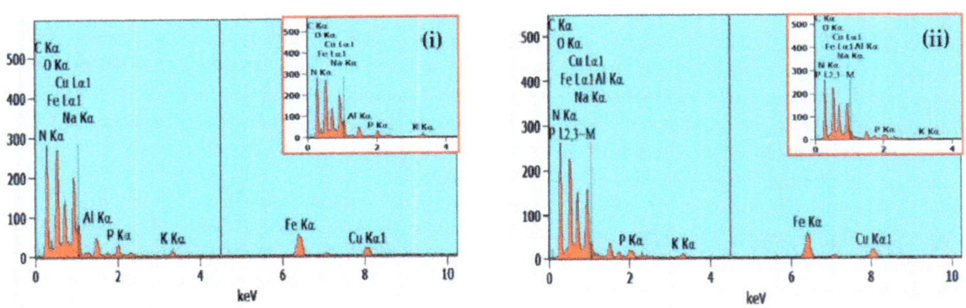

Figure 5. EDX spectra of (**A**)—iron oxide; (**B**)—copper-based TEM grid; (**C**)—MS-1 bacterial cell on two magnetosome crystals (**i**) and (**ii**) along with percentage of each element identified indicated in the table corresponding to each analyzed section.

For a better understanding, respectively for the systematic evaluation of the composition and elemental distribution of MTB, we performed a comparative analysis between the composition of synthetic iron oxide crystals, the TEM grid, and the MS-1 magnetosomes (Figure 5A, Figure 5B, and Figure 5C, respectively).

https://doi.org/10.3390/molecules31020253

Typically, the analysis performed on isolated crystals or on magnetite aggregates shows a ratio of approximately 70 wt% Fe and 30 wt% O, values that approach the theoretical composition of Fe_3O_4, where Fe represents 72.3 wt% and O 27.7 wt% [23]. Our EDX measurements on magnetite crystals confirm the expected elemental composition (Fe and O) while also showing signals from the substrate (Figures 4 and 5A). On the TEM grid (copper with thin carbon film, Figure 5B), the spectra display peaks of Cu and C, as well as elements from the culture medium (Zn, Mg, P). These background contributions must be considered, as accurate identification of the sample's elemental composition relies on careful correlation and interpretation of all detected signals.

In the points analyzed directly on the MS-1 magnetosomes (points 1 and 2—Figure 5C), the EDX spectra revealed a preponderance of iron and oxygen, confirming the nature of the nanoparticles, corresponding to magnetite. The presence of the weak phosphorus signal in these regions is associated with the magnetosome membrane, composed of phospholipids and proteins specific to the biomineralization process. In addition, the residual signals (C, Na, Cu, N, Al, K) can be attributed to the bacterial organic matrix.

EDX analysis of the non-magnetosome regions near magnetotactic bacteria, shown in Figure 5B, shows that oxygen is present in a fairly high proportion, approximately 43% atomic, while iron appears only in very small quantities, around 1.8%. In contrast, the magnetosomal regions analyzed in points 1 and 2 in Figure 5C show a completely different profile: oxygen remains high, around 36%, but iron is much more concentrated, reaching 6–9% atomic. This contrast clearly highlights that magnetosomes are the place where iron accumulates, while the surrounding areas contain only traces, supporting the idea that magnetosome formation involves a localized concentration of Fe and O, specific to magnetite. Of course, the values are not perfectly uniform (taking into account the presence of other elements), which probably reflects natural variations within the cell and/or the limitations of the EDX technique, but the trend is clear and consistent with the literature.

A study by Amor et al. [24] indicates that, although a significant proportion of cellular iron is concentrated as magnetite crystals within magnetosomes (approx. 30%), a notable fraction remains diffusely distributed in the cytoplasm or as a mobile pool of (reducible) Fe(II) and protein-bound complexes. This indicates the existence of intermediate forms or storage forms of iron involved in the controlled biomineralization process. The constant iron EDX signal in the measured points on MS-1 magnetosomes emphasizes the strict control of the biomineralization conducted by the bacteria, and the homogeneous distribution of iron. All these findings confirm the data in the literature regarding the partial storage of iron in the composition of magnetosomes and its diffuse presence in the cytoplasm [24].

Overall, the combined SEM/STEM-EDX analysis confirms that the magnetosomes produced by MS-1 are predominantly composed of iron oxides, their ordered organization in chains being a consequence of the rigorous biological control exercised during biomineralization. The elemental distribution on MS-1 reflects the precise compartmentalization of the biochemical processes involved in the formation and stabilization of these nanocrystals. The EDX method thus offers a powerful combination of structural and chemical analysis and contributes to the understanding of how MS-1 magnetotactic bacteria internally organize their magnetosomes to achieve optimal magnetotactic behavior.

2.3. *AFM Analysis of a Copper TEM Grid*

2.3.1. Untreated with MS-1 Bacterial Cells

AFM images in height sensor mode obtained on the untreated copper TEM grid, as shown in Figure 6, at different scan sizes (1 μm, 3 μm, and 5 μm), highlight the characteristic surface morphology of the support. The topography shows a rough surface (average RMS value of 5.09 nm, 4.31 nm, 8.36 nm at 1 μm, 3 μm, and 5 μm, respectively). Thus, the asperi-

https://doi.org/10.3390/molecules31

ties are relatively homogeneously distributed, associated with the natural microtexture of the carbon film and the granulation of the copper layer (copper grain diameter of 58.7 nm, 121.6 nm, 203.7 nm at 1 µm, 3 µm, and 5 µm, respectively). Analysis of the profile sections (from left side to right side) shows height variations of the order of a few nanometers up to a few hundred nanometers, which confirms the absence of macroscopic contaminants and residual deposits (Figure 6).

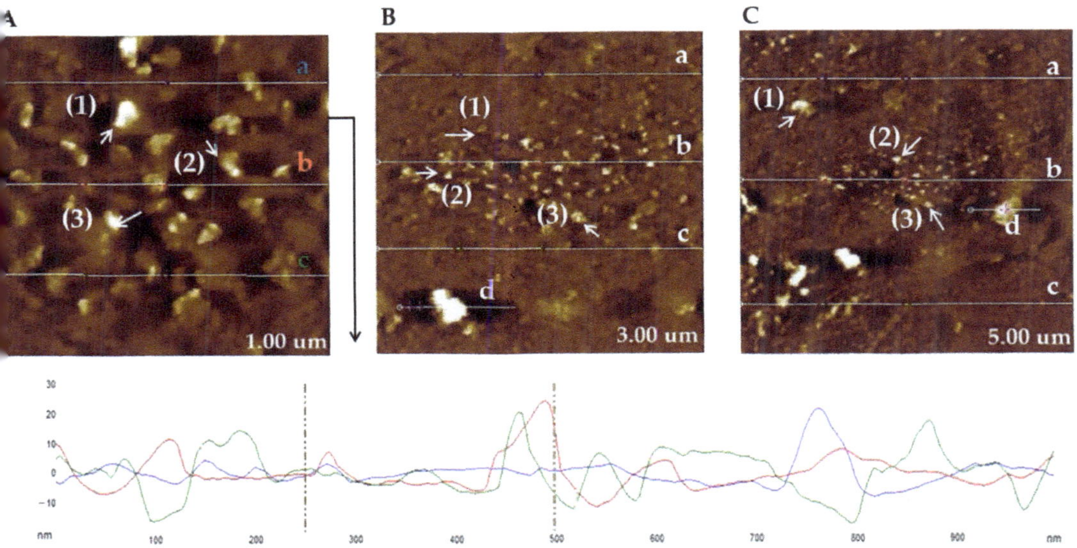

Figure 6. AFM measurements of untreated copper-based TEM grid in the intersection regions between meshes, images of the surface measured of the TEM grid at 1.00 µm (**A**), 3.00 µm (**B**), and 5.00 µm (**C**) (the particle size corresponding to the grid is marked with Arabic numerals, and the sections corresponding to the RMS are marked by letters). Note: AFM measurements indicate that, for the 1 µm grid, the RMS of the analyzed sections (a–c) varies between 2.069, 4.868, and 8.362 nm (the colored letters (a–c) correspond to the colored lines in the image below), while the copper grain particle sizes measured in regions 1–3 are 57.6, 70.1, and 48.6 nm, respectively. In the case of the 3 µm grid, the RMS values (a–d) range from 2.54, 7.69, 2.72, and 7.65 nm across the investigated sections, whereas the particle sizes (1–3) vary between 83.7, 134.8, and 146.5 nm. For the 5 µm grid, a higher surface heterogeneity is observed, with RMS values (a–d) between 4.92, 6.90, 13.27, and 7.98 nm, accompanied by significantly larger particle sizes (1–3) of 314.7, 143.8, and 152.8 nm. RMS—Root Mean Square.

The AFM measurements reveal the topography without prominent foreign particles, confirming that the grid is untreated and suitable for use as a reference support in subsequent real sample experiments. These results establish a control profile of surface roughness and morphology, which can be used for comparison with samples loaded with MS-1 magnetotactic bacteria and their magnetosomes.

2.3.2. Treated with MS-1 Bacterial Cells

To obtain a detailed insight into the morphology and topography of the surface of bacteria, AFM analyses were performed on MS-1. The images obtained provided a faithful representation of the cellular structure and highlight the specific organization of these magnetotactic microorganisms. In Figure 7, we present the topography (in height sensor mode—Figure 7A) and peak force error channel images (Figure 7B), both acquired simultaneously during the scan. The height sensor images provide real height values, but

https://doi.org/10.3390/molecules31020253

the visual representation may lose clarity in areas where the geometry of the bacteria is highly distorted. On the other hand, the Peak Force Error signal highlights rapid changes in relief, producing a light–shadow contrast that highlights fine surface details, difficult to distinguish from the topography alone. It is important to emphasize that peak force error has an exclusively illustrative role, not being suitable for morphometric quantifications, for which only the images obtained in height sensor are used [25–28].

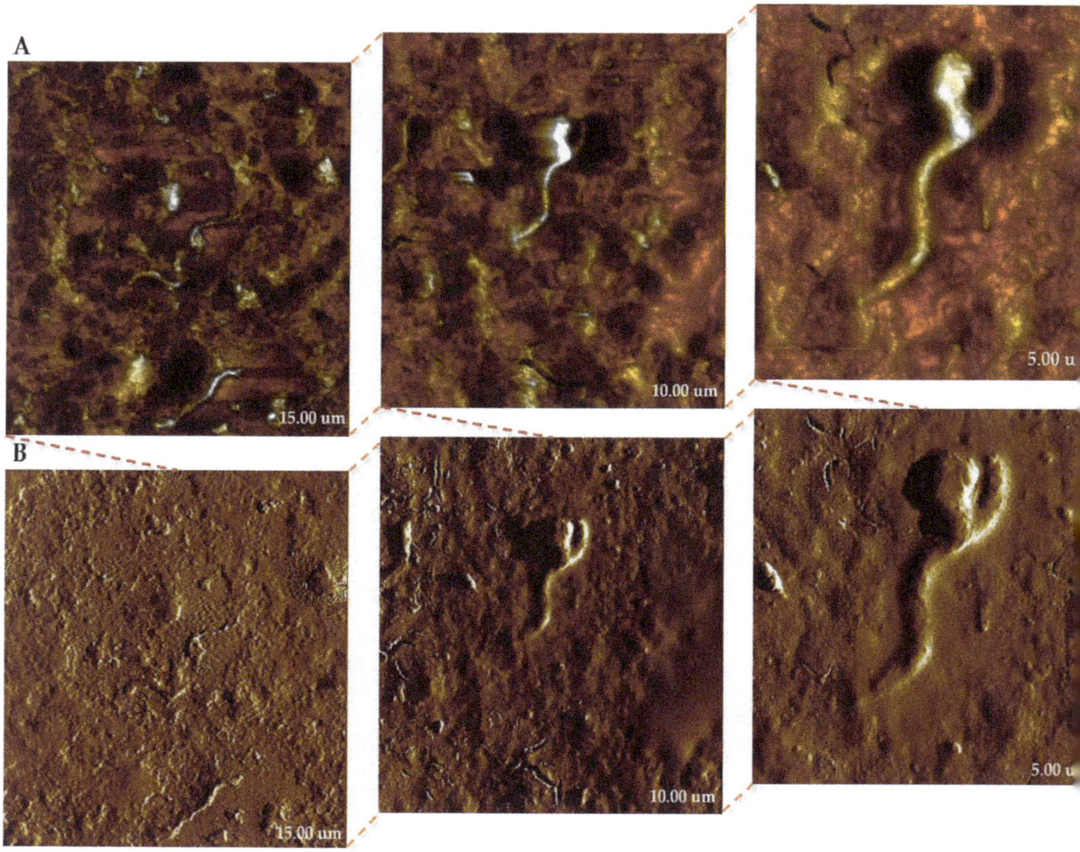

Figure 7. AFM images of MS-1 strain of MTB on copper-based TEM grid at different scales (15, 10, and 5 µm, respectively); (**A**)—height sensor images; (**B**)—peak force error images.

The AFM image obtained on MS-1 (Figure 8) highlights the specific topography of an elongated cell, with spiral morphology, characteristic of the genus Magnetospirillum. The cell presents a continuous, homogeneous surface, without obvious discontinuities, which indicates good structural preservation during the analysis of the sample. The topographic analysis allows the estimation of a length of the bacterium of approximately 3.78 µm, a value consistent with the data obtained by analyzing the bacterium using SEM and STEM echniques.

The topographic contrast reveals fine variations in the surface relief, attributed to the internal chain of magnetosomes aligned along the longitudinal axis, which determine small changes in the locally measured height. Thus, the average roughness values obtained for the bacterial surface indicate a slightly wavy texture, associated with the internal distribution of magnetosomes and ultrastructural variations of the cell membrane (values

https://doi.org/10.3390/molecules310

of 11.20 nm and 13.64 nm at the bacterial ends, and 21.15 nm in the central portion, respectively). In comparison, the RMS values of the background matrix, representing the solidified medium in which the bacteria are suspended, are smaller and relatively close (13.53 nm and 11.97 nm), which confirms the clear delimitation between the bacterial surface and the supporting substrate. This difference between the roughness of the bacteria and that of the matrix highlights the fidelity of the AFM measurements and allows the correlation of the observed morphological details with the internal structure specific to magnetotactic bacteria.

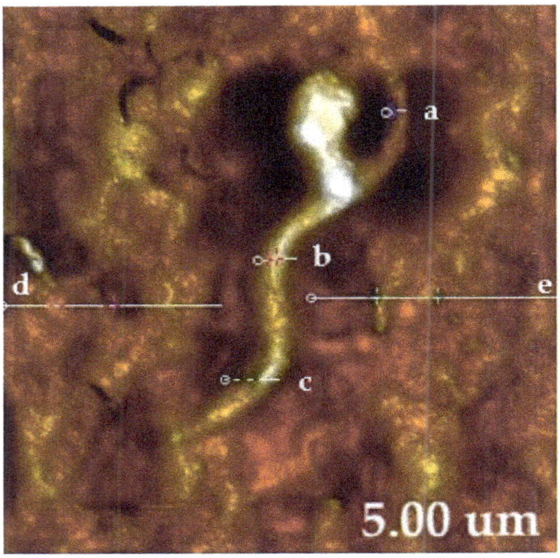

Figure 8. AFM analysis of copper-based TEM grid treated with MS-1 (bacteria size 3.78 μm). Note: For the MS-1 cell section (5 μm), the RMS values measured at three points (at the ends and central portion of the bacteria: (a–c) are 11.20, 21.15, and 13.64 nm, respectively. In the surrounding background matrix, the RMS measured in regions (d,e) is 13.53 and 11.97 nm. RMS refers to root mean square roughness.

3. Discussion

In this study, a long-term stored bacterial suspension was used instead of a freshly isolated one to evaluate how well MS-1 bacteria retain their structural and magnetic properties over time. This approach provided us with essential information about their long-term functional stability and potential relevance in practical applications, where resistance to different environmental conditions is critical.

Sample preparation typically involves fixation, dehydration, and coating with a conductive layer (gold, or carbon), procedures that can alter the native structure of the cells and modify the surface appearance, generating morphological artifacts that are difficult to distinguish from real features [29–32]. Also, because MTB are microaerophilic or strict anaerobic bacteria, their exposure to oxygen during sample preparation can affect cell viability and the integrity of intracellular compounds, including magnetosomes that are sensitive to environmental chemical conditions [29,31,33–35]. The authors proposed the standalone analysis of the MS-1 strain of MTB, without prior coating with a conductive layer. The reasoning and motivation in this regard stemmed from the observation of the limited number of publications performing SEM characterization of MTB strains [36–38], especially on the MS-1 strain [15,39,40].

https://doi.org/10.3390/molecules31020253

Analysis of the MS-1 strain using advanced microscopy (SEM, STEM, EDX, and AFM) provided comprehensive insight into the morphology, structure, and chemical composition of the bacteria. Each method provided complementary information, but most importantly, it confirmed the feasibility of the low-cost TEM metal substrate used in these experiments. SEM provided information on the external morphology, shape, and distribution of magnetosomes [29,41,42] without using a conductive coating, which allowed the bacteria to be observed as close to their natural state as possible. STEM complemented these data by visualizing the internal structure and alignment of the magnetosome chains [43–45]. EDX confirmed the chemical composition of the magnetosomes and highlighted the difference between the signals coming from the substrate and those from the bacteria [46–52]. AFM was useful for analyzing the topography of the substrate and the deposited bacteria, demonstrating the efficient attachment of the cells and the real differences from the substrate background [53,54].

The combined results show that copper-based TEM supports can be used for MS-1 characterization, with certain limitations. SEM, STEM, and EDX allow faithful reproduction of the investigated features, but protocols must be adapted to experimental conditions. AFM can be used, but the presence of substrate roughness can hinder topographic analysis.

Overall, the study highlighted both the advantages and limitations of using this type of low-cost substrate in the microscopic analysis of marine magnetotactic bacteria, emphasizing the novelty of the work: the identification and characterization of bacteria in a state as close as possible to their natural state, using an accessible and reproducible method.

4. Materials and Methods

4.1. Bacterial Culture

Paramagnetospirillum magnetotacticum MS-1 (DSMZ:DSM 3856) grown in standard culture medium provided by the supplier contains 100 µM of Fe(III) quinate solution, seven-vitamin solution, and mineral solution [55,56]. We used an aged bacterial suspension (10 months at 4° C), not a fresh or newly isolated one, in order to investigate and highlight their stability over time.

4.2. Support and Sample Preparation

A drop of MS-1 bacterial suspension (equivalent to 5 µL) was added to the assay support, a copper-based TEM grid, 200 mesh grids (pitch 127 µm, hole width 90 µm, bar width 37 µm) (Ted Pella. Redding, CA, USA). In order to prevent possible contamination when handling, the TEM grid was sampled using a sterile loop. Initially, a 10 µL loop (3 mm diameter) was used. However, due to the larger diameter of the loop, it was adjusted using a 1 µL loop (diameter of 0.5 mm); see Figure 9.

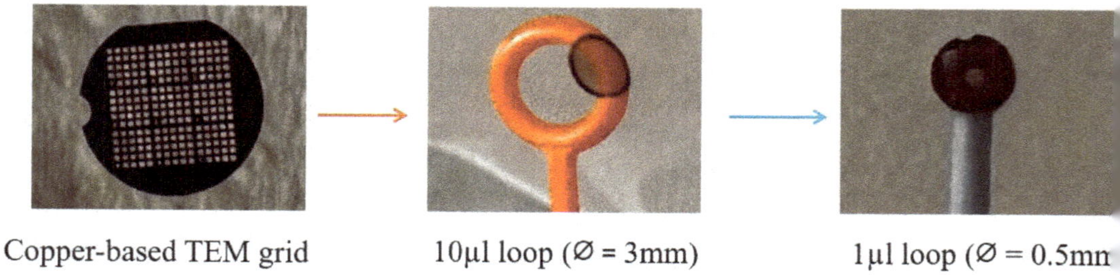

Figure 9. Copper-based TEM grid manipulation.

https://doi.org/10.3390/molecules31

4.3. Microscopic Characterization

MS-1 bacteria were directly adsorbed onto a copper-based TEM grid for SEM, STEM, and EDX analyses. Imaging was performed using a Hitachi FEG SU8030 scanning electron microscope (Hitachi High-Technologies Corporation, Tokyo, Japan) under vacuum conditions ranging from 5 to 25 kV, with a working distance of 10–15 mm. AFM analysis was performed on bacteria adsorbed on copper-grid TEM. Images were obtained with a Bruker, Dimension Icon atomic force microscope (Bruker Corporation, Billerica, MA, USA). The measurements were carried out in air, in ScanAsyst mode, using a silicon tip on a nitride lever and a cantilever with k of 0.4 N/m, f0 of 70 kHz.

5. Conclusions

The integrated application of advanced microscopic techniques (SEM, STEM, EDX, and AFM) on the strain *Paragmnetospirillum magnetotacticum* MS-1 allowed obtaining a complex and complementary picture of the morphology, composition, and organization of bacteria and their magnetosomes. SEM and STEM provided detailed information on the structure of the cells and the arrangement of the magnetosomal chains, EDX confirmed the presence of key elements (especially Fe and O) associated with magnetite crystals, and AFM highlighted the dimension and contour of the cell at the nanometric scale, completing the structural data. The results support the idea of a controlled biomineralization process, characteristic of MTB, and also reveal the existence of intracellular fractions of non-crystallized iron, with relevance for understanding bacterial metabolism.

The use of low-cost copper-based TEM grid support has proven relatively effective for cross-analysis by all four techniques, representing an accessible and reproducible solution for morphological, structural, and elemental studies, with the condition that a rigorous analysis of the signals coming from the substrate is performed.

Beyond the methodological value, these advanced analyses contribute to the consolidation of the role of MTB as sensitive bioindicators and functional nanostructured materials, given their sensitivity to variations in the redox environment and their ability to reflect changes in iron and oxygen concentrations. Thus, the use of several advanced microscopy techniques for characterizing MS-1 not only expands the understanding of the biological mechanisms of biomineralization but also provides valuable information for biomedical and biosensors applications, as well as for ecological surveillance and assessment of the health of ecosystems.

Author Contributions: Conceptualization, R.E.I.; methodology, N.L.P., R.D., J.B., C.O.P., and R.E.I.; software, N.L.P.; validation, R.E.I. and C.O.P.; formal analysis, N.L.P., R.E.I., and C.O.P.; investigation, N.L.P., R.E.I., and C.O.P.; resources, N.L.P., R.E.I., and C.O.P.; data curation, R.E.I. and C.O.P.; writing—original draft preparation, N.L.P.; writing—review and editing—R.E.I.; visualization, N.L.P., R.D., J.B., R.E.I., and C.O.P.; supervision, R.E.I. and C.O.P.; project administration, R.E.I. and C.O.P. All authors have read and agreed to the published version of the manuscript.

Funding: This research received no external funding.

Institutional Review Board Statement: Not applicable.

Informed Consent Statement: Not applicable.

Data Availability Statement: Data are contained within the article.

Acknowledgments: N.L. Paul, a third-year doctoral student, thanks the Erasmus + office, key action 131 for financial support at UTT, France, from January to September 2025.

Conflicts of Interest: The authors declare no conflicts of interest.

https://doi.org/10.3390/molecules31020253

References

1. Bazylinski, D.A.; Frankel, R.B. Magnetosome Formation in Prokaryotes. *Nat. Rev. Microbiol.* **2004**, *2*, 217–230. [Cross [PubMed]
2. Bazylinski, D.A.; Lefèvre, C.T. Magnetotactic Bacteria from Extreme Environments. *Life* **2013**, *3*, 295–307. [CrossRef] [PubM
3. Satyanarayana, S.; Padmaprahlada, S.; Chitradurga, R.; Bhattacharya, S. Orientational Dynamics of Magnetotactic Bacter Earth's Magnetic Field—A Simulation Study. *J. Biol. Phys.* **2021**, *47*, 79–93. [CrossRef] [PubMed]
4. Blakemore, R. Magnetotactic Bacteria. *Science* **1975**, *190*, 377–379. [CrossRef]
5. Paul, N.L.; Carpa, R.; Ionescu, R.E.; Popa, C.O. The Biomedical Limitations of Magnetic Nanoparticles and a Biocompa Alternative in the Form of Magnetotactic Bacteria. *J. Funct. Biomater.* **2025**, *16*, 231. [CrossRef]
6. Faivre, D.; Schüler, D. Magnetotactic Bacteria and Magnetosomes. *Chem. Rev.* **2008**, *108*, 4875–4898. [CrossRef]
7. Bazylinski, D.A.; Garratt-Reed, A.J.; Frankel, R.B. Electron Microscopic Studies of Magnetosomes in Magnetotactic Bac *Microsc. Res. Tech.* **1994**, *27*, 389–401. [CrossRef]
8. Paul, N.L.; Popa, C.O.; Ionescu, R.E. Use of Natural Magnetosome Crystals from Magnetotactic Bacteria for Local Therapy Ve Magnetic Nanoparticles of Similar Compositions, Sizes and Shapes. In *Proceedings of the Selected Papers from ICIR EUROINVE 2025*; Sandu, A.V., Vizureanu, P., Abdullah, M.M.A.B., Nabialek, M., Zainol, M.R.R.M.A., Sandu, I., Eds.; Springer Nature: C Switzerland, 2025; pp. 15–27.
9. Araujo, A.C.V.; Abreu, F.; Silva, K.T.; Bazylinski, D.A.; Lins, U. Magnetotactic Bacteria as Potential Sources of Bioproducts. *Drugs* **2015**, *13*, 389–430. [CrossRef]
10. Paul, N.L.; Popa, C.O.; Ionescu, R.E. Natural Iron Oxide Nanoparticles Produced by Aquatic Magnetotactic Bacteria as I Nanozymes for Nano-Guided Biosensing Platforms—A Systematic Review. *Biosensors* **2025**, *15*, 590. [CrossRef]
11. Ren, G.; Zhou, X.; Long, R.; Xie, M.; Kankala, R.K.; Wang, S.; Zhang, Y.S.; Liu, Y. Biomedical Applications of Magnetosomes: of the Art and Perspectives. *Bioact. Mater.* **2023**, *28*, 27. [CrossRef]
12. Alphandéry, E. Applications of Magnetosomes Synthesized by Magnetotactic Bacteria in Medicine. *Front. Bioeng. Biotechnol.* *2*, 5. [CrossRef] [PubMed]
13. Bertani, L.E.; Weko, J.; Phillips, K.V.; Gray, R.F.; Kirschvink, J.L. Physical and Genetic Characterization of the Genom *Magnetospirillum Magnetotacticum*, Strain MS-1. *Gene* **2001**, *264*, 257–263. [CrossRef] [PubMed]
14. Khalil, I.S.M.; Misra, S. Control Characteristics of Magnetotactic Bacteria: Magnetospirillum Magnetotacticum Strain MS-1 Magnetospirillum Magneticum Strain AMB-1. *IEEE Trans. Magn.* **2014**, *50*, 1–11. [CrossRef]
15. Kundu, S.; Kale, A.A.; Banpurkar, A.G.; Kulkarni, G.R.; Ogale, S.B. On the Change in Bacterial Size and Magnetosome Feat for *Magnetospirillum Magnetotacticum* (MS-1) under High Concentrations of Zinc and Nickel. *Biomaterials* **2009**, *30*, 4211– [CrossRef]
16. Haripriyaa, M.; Suthindhiran, K. Investigation of Pharmacokinetics and Immunogenicity of Magnetosomes. *Artif. Cells Nan Biotechnol.* **2024**, *52*, 69–83. [CrossRef]
17. Yadav, V.K.; Pramanik, S.; Alghamdi, S.; Atwah, B.; Qusty, N.F.; Babalghith, A.O.; Solanki, V.S.; Agarwal, N.; Gupta Niazi, P.; et al. Therapeutic Innovations in Nanomedicine: Exploring the Potential of Magnetotactic Bacteria and Bact Magnetosomes. *Int. J. Nanomed.* **2025**, *20*, 403–444. [CrossRef]
18. Alonso, J.; Gandía, D.; Jefremovas, E.M.; Gandarias, L.; Villanueva, D.; Lete, N.; Marcano, L.; García-Prieto, A.; Or Barquín, L.F.; et al. Magnetotactic Bacteria: Biorobots for Targeted Therapies and Model Nanomagnetic Systems. In Procee of the 2023 IEEE International Magnetic Conference—Short Papers (INTERMAG Short Papers), Sendai, Japan, 15–19 2023; pp. 1–2.
19. Cowley, J.M. Scanning Transmission Electron Microscopy. In *Handbook of Microscopy for Nanotechnology*; Yao, N., Wang, Z.L., Springer: Boston, MA, USA, 2005; pp. 455–491. ISBN 978-1-4020-8006-7.
20. Agibayeva, A.; Guney, M.; Karaca, F.; Kumisbek, A.; Kim, J.R.; Avcu, E. Analytical Methods for Physicochemical Characteriz and Toxicity Assessment of Atmospheric Particulate Matter: A Review. *Sustainability* **2022**, *14*, 13481. [CrossRef]
21. Mirakovski, D.; Damevska, K.; Simeonovski, V.; Nikolovska, S.; Boev, B.; Petrov, A.; Sijakova Ivanova, T.; Zendelska, A.; H Nikolova, M.; Boev, I.; et al. Use of SEM/EDX Methods for the Analysis of Ambient Particulate Matter Adhering to the Surface. *J. Eur. Acad. Dermatol. Venereol.* **2022**, *36*, 1376–1381. [CrossRef]
22. Li, C.; Tardajos, A.P.; Wang, D.; Choukroun, D.; Van Daele, K.; Breugelmans, T.; Bals, S. A Simple Method to Clean Li Contamination on TEM Grids. *Ultramicroscopy* **2021**, *221*, 113195. [CrossRef]
23. Elizondo-Villarreal, N.; Verástegui-Domínguez, L.; Rodríguez-Batista, R.; Gándara-Martínez, E.; Alcorta-García, A.; Mart Delgado, D.; Rodríguez-Castellanos, E.A.; Vázquez-Rodríguez, F.; Gómez-Rodríguez, C. Green Synthesis of Magnetic Nano cles of Iron Oxide Using Aqueous Extracts of Lemon Peel Waste and Its Application in Anti-Corrosive Coatings. *Materials* *15*, 8328. [CrossRef]
24. Amor, M.; Ceballos, A.; Wan, J.; Simon, C.P.; Aron, A.T.; Chang, C.J.; Hellman, F.; Komeili, A. Magnetotactic Bacteria Accum a Large Pool of Iron Distinct from Their Magnetite Crystals. *Appl. Environ. Microbiol.* **2020**, *86*, e01278-20. [CrossRef] [Pub

https://doi.org/10.3390/molecules310

Dujardin, A.; Wolf, P.D.; Lafont, F.; Dupres, V. Automated Multi-Sample Acquisition and Analysis Using Atomic Force Microscopy for Biomedical Applications. *PLoS ONE* **2019**, *14*, e0213853. [CrossRef] [PubMed]

Gour, S.; Mukherjee, A.; Balani, K.; Dhami, N.K. Atomic Force Microscopic Investigations of Transient Early-Stage Bacterial Adhesion and Antibacterial Activity of Silver and Ceria Modified Bioactive Glass. *J. Mater. Res.* **2024**, *39*, 2415–2430. [CrossRef]

Huang, Q.; Wu, H.; Cai, P.; Fein, J.B.; Chen, W. Atomic Force Microscopy Measurements of Bacterial Adhesion and Biofilm Formation onto Clay-Sized Particles. *Sci. Rep.* **2015**, *5*, 16857. [CrossRef]

Schillers, H.; Rianna, C.; Schäpe, J.; Luque, T.; Doschke, H.; Wälte, M.; Uriarte, J.J.; Campillo, N.; Michanetzis, G.P.A.; Bobrowska, J.; et al. Standardized Nanomechanical Atomic Force Microscopy Procedure (SNAP) for Measuring Soft and Biological Samples. *Sci. Rep.* **2017**, *7*, 5117. [CrossRef]

Stadtländer, C.T.K.-H. Scanning Electron Microscopy and Transmission Electron Microscopy of Mollicutes: Challenges and Opportunities. In *Modern Research and Educational Topics in Microscopy*; Méndez-Vilas, A., Díaz, J., Eds.; FORMATEX: Paris, France, 2007; Volume 1, pp. 122–131. ISBN 978-84-611-9419-3.

Bergmans, L.; Moisiadis, P.; Van Meerbeek, B.; Quirynen, M.; Lambrechts, P. Microscopic Observation of Bacteria: Review Highlighting the Use of Environmental SEM. *Int. Endod. J.* **2005**, *38*, 775–788. [CrossRef]

Paul, N.L.; Popa, C.O.; Ionescu, R.E. Updates on the Advantages and Disadvantages of Microscopic and Spectroscopic Characterization of Magnetotactic Bacteria for Biosensor Applications. *Biosensors* **2025**, *15*, 472. [CrossRef]

Datye, A.; DeLaRiva, A. Scanning Electron Microscopy (SEM). In *Springer Handbook of Advanced Catalyst Characterization*; Wachs, I.E., Bañares, M.A., Eds.; Springer International Publishing: Cham, Switzerland, 2023; pp. 359–380. ISBN 978-3-031-07125-6.

Kaláb, M.; Chabot, D. Conventional Scanning Electron Microscopy of Bacteria. *Infocus Mag.* **2008**, *10*, 42–61. [CrossRef]

Relucenti, M.; Familiari, G.; Donfrancesco, O.; Taurino, M.; Li, X.; Chen, R.; Artini, M.; Papa, R.; Selan, L. Microscopy Methods for Biofilm Imaging: Focus on SEM and VP-SEM Pros and Cons. *Biology* **2021**, *10*, 51. [CrossRef]

Lefèvre, C.T.; Bazylinski, D.A. Ecology, Diversity, and Evolution of Magnetotactic Bacteria. *Microbiol. Mol. Biol. Rev.* **2013**, *77*, 497–526. [CrossRef]

Lefèvre, C.T.; Bennet, M.; Landau, L.; Vach, P.; Pignol, D.; Bazylinski, D.A.; Frankel, R.B.; Klumpp, S.; Faivre, D. Diversity of Magneto-Aerotactic Behaviors and Oxygen Sensing Mechanisms in Cultured Magnetotactic Bacteria. *Biophys. J.* **2014**, *107*, 527–538. [CrossRef]

Flies, C.B.; Jonkers, H.M.; de Beer, D.; Bosselmann, K.; Böttcher, M.E.; Schüler, D. Diversity and Vertical Distribution of Magnetotactic Bacteria along Chemical Gradients in Freshwater Microcosms. *FEMS Microbiol. Ecol.* **2005**, *52*, 185–195. [CrossRef] [PubMed]

Salam, M.A.; Korkmaz, N.; Cycil, L.M.; Hasan, F. Isolation, Microscopic and Magnetotactic Characterization of Magnetospirillum Moscoviense MS-24 from Banjosa Lake, Pakistan. *Biotechnol. Lett.* **2023**, *45*, 967–979. [CrossRef] [PubMed]

Marcano, L.; Orue, I.; Gandia, D.; Gandarias, L.; Weigand, M.; Abrudan, R.M.; García-Prieto, A.; García-Arribas, A.; Muela, A.; Fdez-Gubieda, M.L.; et al. Magnetic Anisotropy of Individual Nanomagnets Embedded in Biological Systems Determined by Axi-Asymmetric X-Ray Transmission Microscopy. *ACS Nano* **2022**, *16*, 7398–7408. [CrossRef] [PubMed]

Schaible, G.A.; Kohtz, A.J.; Cliff, J.; Hatzenpichler, R. Correlative SIP-FISH-Raman-SEM-NanoSIMS Links Identity, Morphology, Biochemistry, and Physiology of Environmental Microbes. *ISME Commun.* **2022**, *2*, 52. [CrossRef]

Khalil, I.S.M.; Pichel, M.P.; Zondervan, L.; Abelmann, L.; Misra, S. Characterization and Control of Biological Microrobots. In *Experimental Robotics: The 13th International Symposium on Experimental Robotics*; Desai, J.P., Dudek, G., Khatib, O., Kumar, V., Eds.; Springer International Publishing: Heidelberg, Germany, 2013; pp. 617–631. ISBN 978-3-319-00065-7.

Khalil, I.S.M.; Magdanz, V.; Sanchez, S.; Schmidt, O.G.; Misra, S. Magnetotactic Bacteria and Microjets: A Comparative Study. In Proceedings of the 2013 IEEE/RSJ International Conference on Intelligent Robots and Systems, Tokyo, Japan, 3–7 November 2013; pp. 2035–2040.

Ilett, M.; S'ari, M.; Freeman, H.; Aslam, Z.; Koniuch, N.; Afzali, M.; Cattle, J.; Hooley, R.; Roncal-Herrero, T.; Collins, S.M.; et al. Analysis of Complex, Beam-Sensitive Materials by Transmission Electron Microscopy and Associated Techniques. *Philos. Trans. R. Soc. A Math. Phys. Eng. Sci.* **2020**, *378*, 20190601. [CrossRef]

Ophus, C. Quantitative Scanning Transmission Electron Microscopy for Materials Science: Imaging, Diffraction, Spectroscopy, and Tomography. *Annu. Rev. Mater. Res.* **2023**, *53*, 105–141. [CrossRef]

Vahidi, H.; Syed, K.; Guo, H.; Wang, X.; Wardini, J.L.; Martinez, J.; Bowman, W.J. A Review of Grain Boundary and Heterointerface Characterization in Polycrystalline Oxides by (Scanning) Transmission Electron Microscopy. *Crystals* **2021**, *11*, 878. [CrossRef]

Davies, T.E.; Li, H.; Bessette, S.; Gauvin, R.; Patience, G.S.; Dummer, N.F. Experimental Methods in Chemical Engineering: Scanning Electron Microscopy and X-Ray Ultra-Microscopy—SEM and XuM. *Can. J. Chem. Eng.* **2022**, *100*, 3145–3159. [CrossRef]

Hodoroaba, V.-D. Chapter 4.4—Energy-Dispersive X-Ray Spectroscopy (EDS). In *Characterization of Nanoparticles*; Hodoroaba, V.-D., Unger, W.E.S., Shard, A.G., Eds.; Micro and Nano Technologies; Elsevier: Amsterdam, The Netherlands, 2020; pp. 397–417. ISBN 978-0-12-814182-3.

https://doi.org/10.3390/molecules31020253

48. Haddad, G.; Takakura, T.; Bellali, S.; Fontanini, A.; Ominami, Y.; Khalil, J.B.; Raoult, D. A Preliminary Investigation into Bac Viability Using Scanning Electron Microscopy–Energy-Dispersive X-Ray Analysis: The Case of Antibiotics. *Front. Microbiol.* 13, 967904. [CrossRef]
49. Khan, M.S.I.; Oh, S.-W.; Kim, Y.-J. Power of Scanning Electron Microscopy and Energy Dispersive X-Ray Analysis in F Microbial Detection and Identification at the Single Cell Level. *Sci. Rep.* **2020**, *10*, 2368. [CrossRef] [PubMed]
50. Rimal, B.; Chang, J.D.; Liu, C.; Kim, H.; Aderotoye, O.; Zechmann, B.; Kim, S.J. Scanning Electron Microscopy and En Dispersive X-Ray Spectroscopy of Staphylococcus Aureus Biofilms. *ACS Omega* **2024**, *9*, 37610–37620. [CrossRef] [PubMed
51. Ashfaq, M.Y.; Al-Ghouti, M.A.; Al Disi, Z.A.; Zouari, N. Investigating the Microorganisms-Calcium Sulfate Interaction in Re Osmosis Systems Using SEM-EDX Technique. *J. Environ. Chem. Eng.* **2020**, *8*, 103963. [CrossRef]
52. Verma, J.; Khanna, A.S.; Sahney, R.; Bhattacharya, A. Super Protective Anti-Bacterial Coating Development with Silica–Ti Nano Core–Shells. *Nanoscale Adv.* **2020**, *2*, 4093–4105. [CrossRef]
53. Oestreicher, Z.; Valverde-Tercedor, C.; Chen, L.; Jimenez-Lopez, C.; Bazylinski, D.A.; Casillas-Ituarte, N.N.; Lower, Lower, B.H. Magnetosomes and Magnetite Crystals Produced by Magnetotactic Bacteria as Resolved by Atomic Force Micro and Transmission Electron Microscopy. *Micron* **2012**, *43*, 1331–1335. [CrossRef]
54. Marqués-Marchán, J.; Jaafar, M.; Ares, P.; Gubieda, A.G.; Berganza, E.; Abad, A.; Fdez-Gubieda, M.L.; Asenjo, A. Mag Imaging of Individual Magnetosome Chains in Magnetotactic Bacteria. *Biomater. Adv.* **2024**, *163*, 213969. [CrossRef]
55. Leibniz Institute DSMZ: Details. Available online: https://www.dsmz.de/collection/catalogue/details/culture/DSM (accessed on 27 July 2025).
56. Balkwill, D.L.; Maratea, D.; Blakemore, R.P. Ultrastructure of a Magnetotactic Spirillum. *J. Bacteriol.* **1980**, *141*, 1399. [Cross

molecules

MDPI

Article

Au Nanoshell-Based Lateral Flow Immunoassay for Colorimetric and Photothermal Dual-Mode Detection of Interleukin-6

Chongying Wen [1,*], Yue Dou [1], Yao Liu [1], Xuan Jiang [1], Xiaomei Tu [1] and Ruiqiao Zhang [2,*]

1 College of Chemistry and Chemical Engineering, China University of Petroleum (East China), Qingdao 266580, China; douy126@126.com (Y.D.); 17663987932@163.com (Y.L.); 17660637078@163.com (X.J.); 2203040221@s.upc.edu.cn (X.T.)
2 Qingdao Academy of Agricultural Sciences, Qingdao 266100, China
* Correspondence: flcyxt@163.com (C.W.); aqiaohappy@163.com (R.Z.)

Abstract: Interleukin-6 (IL-6) detection and monitoring are of great significance for evaluating the progression of many diseases and their therapeutic efficacy. Lateral flow immunoassay (LFIA) is one of the most promising point-of-care testing (POCT) methods, yet suffers from low sensitivity and poor quantitative ability, which greatly limits its application in IL-6 detection. Hence, in this work, we integrated Au_{shell} nanoparticles (NPs) as new LFIA reporters and achieved the colorimetric and photothermal dual-mode detection of IL-6. Au_{shell} NPs were conveniently prepared using a galvanic exchange process. By controlling the shell thickness, their localized surface plasmon resonance (LSPR) peak was easily tuned to near-infrared (NIR) range, which matched well with the NIR irradiation light. Thus, the Au_{shell} NPs were endowed with good photothermal effect. Au_{shell} NPs were then modified with IL-6 detection antibody to construct Au_{shell} probes. In the LFIA detection, the Au_{shell} probes were combined with IL-6, which were further captured by the capture IL-6 antibody on the test line of the strip, forming a colored band. By observation with naked eyes, the colorimetric qualitative detection of IL-6 was achieved with limit of 5 ng/mL. By measuring the temperature rise of the test line with a portable infrared thermal camera, the photothermal quantitative detection of IL-6 was performed from 1~1000 ng/mL. The photothermal detection limit reached 0.3 ng/mL, which was reduced by nearly 20 times compared with naked-eye detection. Therefore, this Au_{shell}-based LFIA efficiently improved the sensitivity and quantitative ability of commercial colloidal gold LFIA. Furthermore, this method showed good specificity, and kept the advantages of convenience, speed, cost-effectiveness, and portability. Therefore, this Au_{shell}-based LFIA exhibits practical application potential in IL-6 POCT detection.

Keywords: Au_{shell} nanoparticles; Interleukin-6; lateral flow immunoassay; colorimetric detection; photothermal detection

1. Introduction

Interleukin-6 (IL-6) is an important pleiotropic cytokine, which is closely related to immune regulation, inflammation, hematopoiesis, oncogenesis, and so on [1,2]. The IL-6 level increases in many diseases, such as infectious diseases, immunological diseases, multiple sclerosis, Alzheimer disease, and various cancers [3–5]. The increase degree is positively correlated to the severity of disease [6–8]. Now, IL-6 detection and monitoring have been used to evaluate the progression of many diseases and their therapeutic efficacy. Currently, clinical detection methods including enzyme linked immunoassay, radioimmunoassay, and chemiluminescent immunoassay have high sensitivity and accuracy, yet suffer from the drawbacks of time-consuming and complicated manipulation, sophisticated equipment, and a high dependence on professional staff and workplaces [9–11]. With the increasing demand for point-of-care testing (POCT) and home self-testing, it is very necessary to develop simple, rapid, sensitive, and low-cost methods for detecting IL-6.

Lateral flow immunoassay (LFIA) possesses the characteristics of convenience, portability, rapidity, and cost-effectiveness, which does not require complicated instruments or skilled staff [12–14]. Thus, LFIA shows promising advantages in POCT. As a successful example, pregnancy test strips have been applied by numerous people to achieve rapid home self-test, avoiding the trouble of going to hospital. However, nearly all of the commercial LFIAs use gold nanoparticles (NPs) as reporters. These colloidal gold strips can only give a "yes or no" result and cannot achieve accurate quantification. Moreover, due to the limited localized surface plasmon resonance (LSPR) effect of gold NPs, colloidal gold strips exhibit low sensitivity [15–17]. These limitations greatly restrict their application to IL-6 detection.

Recently, researchers have made various efforts to improve the performance of traditional LFIA. On the one hand, researchers have developed various signal amplification methods to enhance the colorimetric signals on the test zones, such as enzyme-based color enhancement and in situ noble metal shell growth [18–21]. These methods indeed increase the sensitivity for naked-eye determination. However, the additional amplification steps usually make the manipulation more complex and time-consuming. On the other hand, some researchers have modified the LFIA and developed a vertical flow assay (VFA) [22,23]. VFA can process a greater volume of samples or high-viscosity samples, thus improving the detection sensitivity. But VFAs usually have complicated operation principles, making them unable to compete with LFIAs in practice [24]. Compared with the first two attempts, integrating new nanomaterials with better signals into a LFIA to replace gold NPs may be a more convenient and effective method to improve the limitations of traditional LFIAs. Currently, quantum dots, surface-enhanced Raman scattering nanomaterials, magnetic nanomaterials, photothermal nanomaterials, and so on have been used in LFIAs, achieving higher sensitivity and accurate quantification [25–29]. Among all these new labels, photothermal nanomaterials can convert light into heat, producing temperature rises as signals, which do not need a complex or expensive readout instrument [30–32]. Moreover, the development of highly sensitive infrared cameras which can distinguish ± 0.025 °C makes photothermal sensing usually have a high resolution [33,34]. Furthermore, the temperature signal is hardly interfered by the matrix color, endowing photothermal detection with a strong anti-interference ability [34,35]. Thus, photothermal nanomaterials have become one of the most promising LFIA reporters. For example, Zhang et al. constructed a black phosphorus-Au nanocomposite to function as a LFIA signal probe and achieved an ultrasensitive detection of 17β-estradiol and mycotoxin zearalenone with limits of 50 pg/mL and 10 pg/mL, respectively, which was nearly 100-fold more sensitive than a visual LFIA [36,37]. As another example, Hao et al. prepared $ReSe_2$ nanosheets using liquid exfoliation, and with them a photothermal LFIA detection of human anti-SARS-CoV-2 S protein antibodies was performed with a detection limit of 0.86 ng/mL, which was 108-fold lower than that of the colorimetric signal [38]. In recent years, our group also has been dedicated to developing photothermal LFIAs. For example, we designed dumbbell-like Au-Fe_3O_4 NPs with a seed-mediated growth method, and realized the sensitive photothermal detection of *Salmonella* and SARS-CoV-2 [39,40]. The above studies show that integrating photothermal nanomaterials as new reporters indeed improves the detection sensitivity of a traditional LFIA. However, most of the photothermal nanomaterials usually comprise two parts: the colorimetric part such as Au and the photothermal part such as black phosphorus. Thus, the preparation procedures of most of the photothermal nanomaterials were much more complicated than gold NPs, limiting their practical application. Therefore, it is necessary to develop a photothermal LFIA reporter with a simple and low-cost synthetic method.

At present, near-infrared (NIR) irradiation light is most commonly used in photothermal detection, which can greatly reduce the background signal. Thus, to obtain a good photothermal effect, it is very necessary to tune the nanomaterials' LSPR absorption peak to the NIR range. As reported, the gold nanoshells' LSPR peak can be easily tuned from the visible to the NIR range just by controlling the shell thickness [41,42]. Hence, in this work, we used a galvanic exchange process to prepare NIR gold nanoshells, and applied them to a dual-mode LFIA detection of IL-6. The gold nanoshells possessed a good colorimetric signal and photothermal effect without combining other photothermal nanomaterials. Furthermore, the synthesis step could be easily performed. Using gold nanoshells as LFIA reporters, by observation with naked eyes, the colorimetric qualitative detection of IL-6 was achieved with a limit of 5 ng/mL. By measuring the temperature change in the test lines (T lines) after laser irradiation, the photothermal quantitative detection of IL-6 was performed from 1~1000 ng/mL. Under optimal conditions, the photothermal detection limit reached 0.3 ng/mL, which was reduced by nearly 20 times compared with that of naked-eye detection. Therefore, this Au_{shell}-based LFIA efficiently improved the sensitivity and quantitative ability of commercial colloidal gold LFIA, showing a practical application value.

2. Results and Discussion

2.1. Principle of the Au_{shell}-Based LFIA for Detecting IL-6

As illustrated in Scheme 1A, based on the antibody–antigen reaction, the Au_{shell} probes combine with IL-6 and then are loaded on the strip; these are then captured by the IL-6 capture antibody on the T line, forming a colored band. Free Au_{shell} probes migrate further and bind with the goat anti-mouse secondary antibody on the control line (C line), forming a second colored band. Thus, positive samples produce two colored bands, while in negative samples, no Au_{shell} probe-IL-6 complexes can accumulate in the T line, and only the Au_{shell} probes are captured by the C line. Thus, negative samples only produce a colored C line. In cases where the C line does not show any color, it means that the antibodies on the probes or the strips are inactivated, and the results are invalid. By observing the color bands with naked eyes, the colorimetric qualitative detection of IL-6 is achieved. By measuring the temperature change in the T lines after laser irradiation, the photothermal quantitative detection of IL-6 is performed. The typical detection results of the positive and negative samples are illustrated in Scheme 1B. It can be seen that the positive sample produced two obvious blue-black bands, while the negative sample only showed one blue-black control band. In the photothermal images, the T line of the positive sample showed an obvious high temperature, while the T line of negative sample did not show any obvious high-temperature zones.

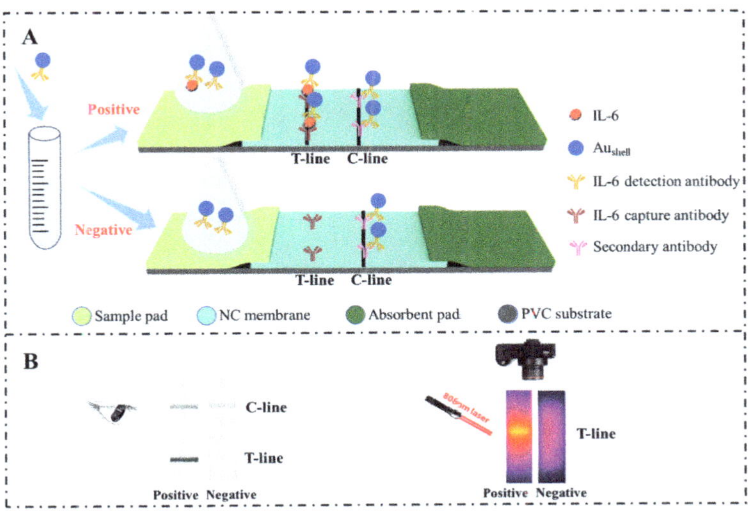

Scheme 1. (**A**) Schematic diagram for the principle of the Au_{shell}-based LFIA for detecting IL-6. (**B**) Typical naked-eye observation and photothermal results of positive and negative samples.

2.2. Characterization of Au_{shell} and Au_{shell} Probes

To obtain the best photothermal effect, the LSPR absorption peak of the Au_{shell} NPs should match the 808 nm irradiation laser. With increasing the amount of chloroauric acid, the etching enhanced, and the Au_{shell} absorption peak red shifted. As shown in Figure 1A, after five additions of chloroauric acid, the absorption peak reached 779 nm. Further increasing the chloroauric acid amount would make the peak shift further to longer wavelength. However, excessive chloroauric acid would destroy the Au_{shell} structure, resulting in gold rings, which would decrease the stability [42]. Hence, through the comprehensive consideration of the photothermal effect and stability, five additions of chloroauric acid were chosen. The obtained Au_{shell} showed satisfactory dispersibility with a diameter of 32.8 nm (Figure 1B,C). After modification with IL-6 detection antibody, their hydrodynamic diameter increased from 46.96 nm to 62.08 nm (Figure 1D). When the Au_{shell} and Au_{shell} probes were loaded on the IL-6 strips, only the Au_{shell} probes could produce a colored C line, while Au_{shell} NPs could not accumulate on the C line (Figure 1E). The above results suggested that the IL-6 detection antibody was successfully conjugated with the Au_{shell} NPs and retained a good bioactivity.

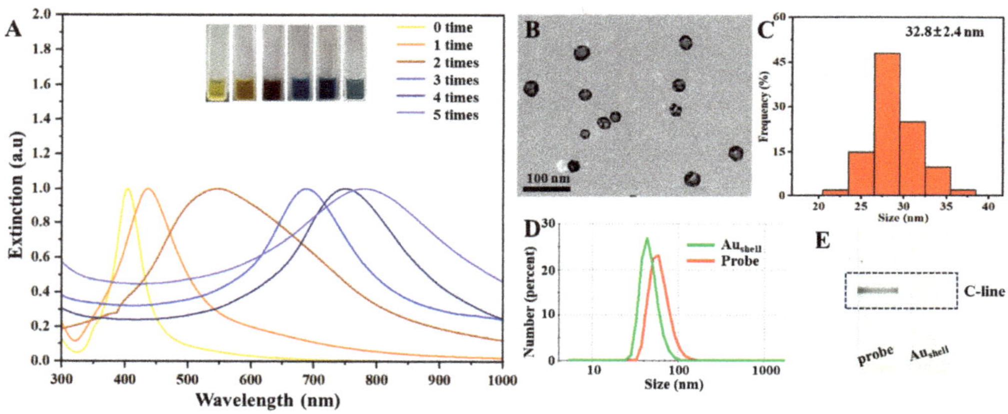

Figure 1. (**A**) UV-Vis spectra and corresponding photos of Au_{shell} NPs with different times of chloroauric acid additions. (**B**) TEM image of Au_{shell} NPs. (**C**) Size distribution of Au_{shell} NPs. (**D**) Hydrodynamic diameter distributions of Au_{shell} NPs and Au_{shell} probes. (**E**) Photos of the IL-6 strips loaded with Au_{shell} NPs and Au_{shell} probes.

2.3. Investigation of the Optimal Detection Conditions

To achieve the best analytical performance, several important detection conditions were optimized, such as irradiation power density and time, bovine serum albumin (BSA) concentration for blocking, Au_{shell} probe concentration, amount of Au_{shell} probes, and incubation time. As shown in Figure 2A, with an increase in the irradiation power density, ΔT increased. When 1.44 W/cm^2 was used, ΔT reached more than 50 °C, and correspondingly, the temperature attained from the tested strip reached as high as 99.5 °C. In consideration of protecting the strips, 1.16 W/cm^2 was chosen as the optimal power density. Similarly, with increasing irradiation time, the temperature of the tested strip increased. At 100 s, the temperature reached a stable and maximum value (Figure 2B), which was chosen as the optimal irradiation time. For other conditions' optimization, the condition which could obtain the highest ratio of the positive sample signal to the negative sample signal (signal-to-noise ratio) was selected. According to the experiment results (Figure 2C–F), 15% BSA, Au_{shell} probes concentrated by five times, 20 μL of Au_{shell} probes, and 10 min incubation were finally chosen for further detection.

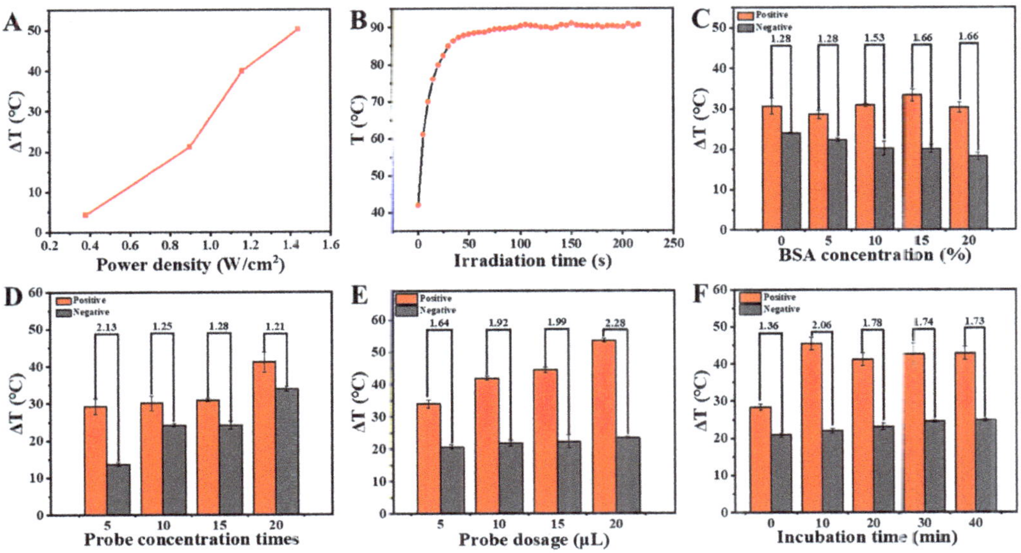

Figure 2. Photothermal signals obtained at different irradiation power densities (**A**), irradiation times (**B**), BSA concentrations for blocking (**C**), Au_{shell} probe concentration times (**D**), Au_{shell} probe amounts (**E**), and incubation times (**F**). The numbers above the columns are the signal-to-noise ratios. Error bars = ±SD (n = 3).

2.4. *Colorimetric Qualitative Detection of IL-6 with the* Au_{shell}*-Based LFIA*

Under optimal detection conditions, we investigated the colorimetric qualitative detection performance of the Au_{shell}-based LFIA. As shown in Figure 3, the negative sample only produced one colored C line. When the IL-6 concentration reached 5 ng/mL, the T line began to exhibit a visible blue-black color, and with the IL-6 concentration further increasing, the color of the T line became darker. This was because more IL-6 would produce more IL-6-Au_{shell} complexes, and hence more Au_{shell} would accumulate on the T line, thus resulting in a darker color. By observing the T line color, the colorimetric qualitative detection of IL-6 could be achieved with a naked-eye detection limit of 5 ng/mL.

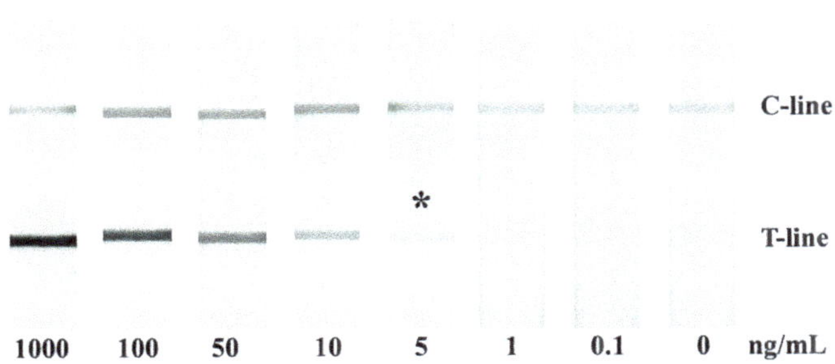

Figure 3. Photos of LFIA strips for detection of different concentrations of IL-6. The asterisk represents the naked-eye limit of detection (LOD).

2.5. *Photothermal Quantitative Detection of IL-6 with the* Au_{shell}*-Based LFIA*

Under optimal detection conditions, we further investigated the photothermal quantitative detection performance of the Au_{shell}-based LFIA. As shown in Figure 4, with the increase in IL-6 concentration, the ΔT of the T line increased. In the IL-6 concentration range of 1~1000 ng/mL, the ΔT was clearly linear with the common logarithm of the IL-6 concentration. The LOD was calculated to be 0.3 ng/mL with the standard IUPAC method (LOD signal = $\Delta T_{blank} + 3SD_{blank}$, where ΔT_{blank} is the average ΔT obtained from 11 blank samples, and SD_{blank} is their standard deviation). It could be seen that the photothermal detection not only achieved the quantification of IL-6, but it also enhanced the detection sensitivity. The photothermal LOD was reduced by nearly 20 times compared with that of naked-eye detection.

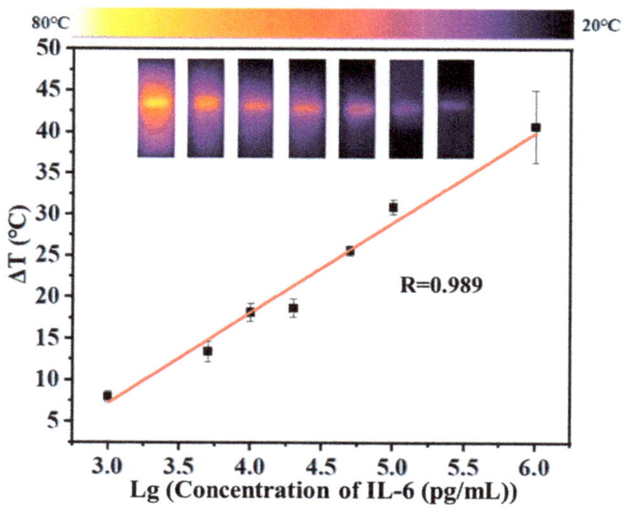

Figure 4. Linear relationship of photothermal signal (ΔT) versus IL-6 concentration. The corresponding photothermal images are shown in the inset. Error bars = ±SD (n = 3).

2.6. Specificity Investigation

To investigate the specificity of the Au_{shell}-based LFIA, several common cytokines with 10 times higher concentrations than IL-6 were detected. As shown in Figure 5A, IL-6 groups showed obvious colored T line and C line, while in other cytokine groups, the T lines were invisible, and only the C line exhibited color. When measuring the photothermal signals, the IL-6 groups showed high ΔT, while the ΔTs from other cytokine groups were very low, which were nearly the same as those from the blank groups (Figure 5B). Hence, this LFIA had good specificity.

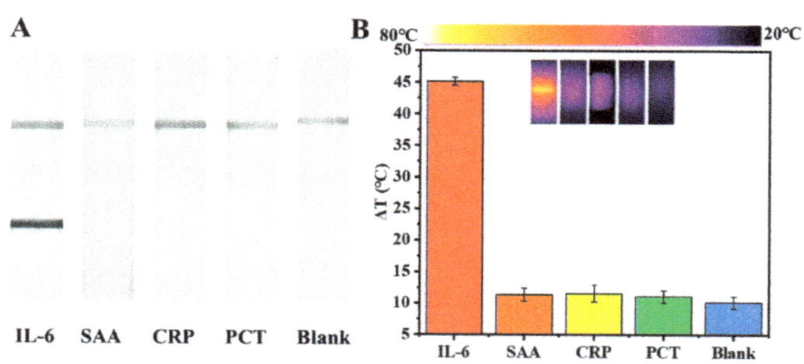

Figure 5. Results from detection of different cytokines with the Au_{shell}-based LFIA. (**A**) Photos of the test strips. (**B**) ΔT histograms and photothermal images of the test strips.

3. Experimental Section

3.1. Reagents and Instruments

Polyvinylpyrrolidone, trisodium citrate, potassium carbonate, sucrose, sodium borohydride, chloroauric acid, Tween-20, and silver nitrate were purchased from Macklin Biochemical Co., Ltd. (Shanghai, China). Bovine serum albumin (BSA) was obtained from Saiguo biotech Co., Ltd. (Guangzhou, China). PBS (pH = 7.4) was purchased from Labgic Technology Co., Ltd. (Beijing, China). IL-6, IL-6 detection antibody, IL-6 capture antibody, goat anti-mouse secondary antibody, serum amyloid protein A (SAA), C-reactive protein (CRP), and procalcitonin (PCT) were supplied by Henderson Biotechnology Co., Ltd. (Qingdao, China). Nitrocellulose (NC) membrane (CN95), sample pad, absorbent pad, and polyvinyl chloride (PVC) substrate were purchased from Joey-biotech Co., Ltd. (Shanghai, China). A UV-Vis spectrophotometer (UV-2450, Shimadzu, Kyoto, Japan) was used to measure absorption spectra. An electron microscope (JEM 1400, JEOL, Tokyo, Japan) was used to take transmission electron microscopy (TEM) images. A precision sprayer (HGS510, AUTOKUN, Hangzhou, China) and high-speed guillotine cutter (HGS210, AUTOKUN, Hangzhou, China) were used to spray and divide test strips. A Zetasizer Nano ZS instrument (Malvern, Britain) was used to measure hydrodynamic diameter. An infrared thermal imager (FOTRIC 226s, Shanghai, China) and an 808 laser (FU808AD2000-F34, Shenzhen, China) were used to measure photothermal signals.

3.2. Preparation of Au Nanoshells

Au nanoshells were prepared via a nanoscale galvanic exchange process according to the previous work [42]. A total of 20 mL of 1% trisodium citrate and 75 mL of ultra-pure water were mixed and heated to 70 °C. After 15 min, 1.7 mL of 1% silver nitrate was added and 2 mL of 1% sodium borohydride was quickly injected for 1 h reaction to obtain Ag seeds. Then, 20 mL of Ag seeds and 3.4 mL of 1% silver nitrate were added to the boiling mixture, containing 4 mL of 1% trisodium citrate and 150 mL of ultra-pure water. The reaction lasted for 1 h. Afterwards, 4 mL of 1% trisodium citrate and 3.4 mL of 1% silver nitrate were added to the above mixture for another 1 h reaction. Trisodium citrate and silver nitrate were added and reacted once again. After cooling down to room temperature, the above solution was irradiated with 254 nm ultraviolet light for 40 min, and Ag NPs were obtained. Then, 5 mL of Ag NPs and 45 mL of polyvinylpyrrolidone (1 mg/mL) were mixed and heated to 100 °C. After 15 min, chloroauric acid was added drop by drop at a rate of 1 mL/min every 2 min. After 5 times, the reaction was terminated by ice water, and Au nanoshells were obtained.

3.3. Construction of Au_{shell} Probes Targeting IL-6

Au nanoshells were first concentrated by centrifugation. Then, 1 mL of the concentrated Au nanoshells was added with 0.1 mol/L potassium carbonate to adjust pH to 8~9, and 29.4 μL of IL-6 detection antibody (1 mg/mL) was added for 45 min reaction with shaking. After that, the Au_{shell} probes were centrifuged and resuspended in BSA solution for 30 min blocking. Finally, the Au_{shell} probes were washed 3 times using centrifugation and resuspended in 1 mL of pH = 7.4 PBS containing 1% BSA, 0.5% tween-20, and 1% sucrose for further use.

3.4. Fabrication of IL-6 LFIA Strips

As illustrated in Scheme 1, the strip was composed of PVC substrate, sample pad, NC membrane, and absorbent pad. For fabrication, NC membrane was first pasted on middle of the PVC substrate. Then, 1.7 cm width absorbent pad was pasted on the right with ca. 0.3 cm overlap, and 2.2 cm width sample pad was pasted on the left with ca. 0.3 cm overlap. T line was sprayed with IL-6 capture antibody (1 mg/mL) and C line was sprayed with goat anti-mouse secondary antibody (1 mg/mL), both at a rate of 1 μL/cm. Finally, the strips were dried at 37 °C and then cut into 3 mm width for use.

3.5. Detection of IL-6 with the Au_{shell}-Based LFIA

Typically, 20 μL of sample solution were added with Au_{shell} probes, which were further added with PBS (pH = 7.4, containing 1% BSA, 0.5% tween-20, and 1% sucrose) to make the total volume reach 50 μL. The above mixture was added to a microplate. After incubation with gentle shaking, the sample pad of the LFIA strip was immersed in the solution. After 15 min, by observing the T line colors with naked eyes, colorimetric qualitative detection of IL-6 was achieved. By measuring temperature change (ΔT) of T lines after 808 nm laser irradiation, photothermal quantitative detection of IL-6 was performed. To eliminate the impact of irradiation on the strip temperature, ΔT was defined as follows:

$$\Delta T = T_1 - T_0 \quad (1)$$

where T_1 and T_0 are, respectively, the temperature of the highest temperature zone attained from the tested strip and unused strip after irradiation.

3.6. Optimization of the Detection Conditions

For irradiation power density optimization, the test line of one strip was irradiated, respectively, at 0.38, 0.9, 1.16, and 1.44 W/cm^2, and the corresponding ΔT was measured for comparison. For the irradiation time optimization, the ΔTs obtained at different irradiation times were recorded for comparison. For the optimization of BSA concentration for blocking, the Au_{shell} probe concentration, the amount of Au_{shell} probes, and the incubation time, a single factor variable method was performed. This involves only varying one condition while fixing other conditions, and the corresponding signal-to-noise ratio was calculated for comparison. Specifically, for BSA concentration optimization, 0%, 5%, 10%, 15%, and 20% BSA were, respectively, used to block the Au_{shell} probes, and applied to the IL-6 LFIA detection under the other fixed conditions (Au_{shell} probes by 10 times concentration, 10 μL of Au_{shell} probes, and 10 min incubation). For Au_{shell} probe concentration optimization, Au_{shell} probes with peak absorbance of ca. 0.7 were, respectively, concentrated by 5, 10, 15, 20 times and then used for IL-6 detection under the other fixed conditions (5% BSA, 10 μL of Au_{shell} probes, and 10 min incubation). For Au_{shell} probe amount optimization, 5, 10, 15, and 20 μL of Au_{shell} probes were, respectively, used under the other fixed conditions (5% BSA, Au_{shell} probes by 5 times concentration, and 10 min incubation). For incubation time optimization, 0, 10, 20, 30, and 40 min were, respectively, used under the other fixed conditions (5% BSA, Au_{shell} probes by 5 times concentration, and 10 μL of Au_{shell} probes).

4. Conclusions

In summary, we developed an Au nanoshell-based LFIA for the colorimetric and photothermal dual-mode detection of IL-6. The Au nanoshells were conveniently prepared by galvanic exchange process. Its LSPR peak could be easily tuned to the NIR range to match well with the NIR irradiation light. As a result, the Au nanoshells possessed a good colorimetric signal and a high photothermal effect without combining with other photothermal nanomaterials. We integrated the Au nanoshells as LFIA reporters and systematically optimized the related detection conditions. Under optimal conditions, the colorimetric qualitative detection of IL-6 could be achieved with a naked-eye LOD of 5 ng/mL, and the photothermal quantitative detection of IL-6 was performed from 1~1000 ng/mL with a LOD of 0.3 ng/mL, which was reduced by nearly 20 times compared with naked-eye detection. Therefore, the photothermal detection not only achieved the quantification of IL-6, but also enhanced the detection sensitivity. Furthermore, this Au nanoshell-based LFIA had good specificity, could be manipulated conveniently and rapidly, and did not need sophisticated readout instrument. Thus, this Au nanoshell-based LFIA efficiently improved the limitations of traditional LFIAs, and showed a practical application potential in IL-6 POCT detection.

Author Contributions: Conceptualization, C.W. and Y.D.; methodology, Y.D. and Y.L.; validation, C.W., Y.D. and Y.L.; formal analysis, Y.D. and Y.L.; investigation, C.W., Y.D. and Y.L.; data curation, X.J. and X.T.; writing—original draft preparation, C.W.; writing—review and editing, C.W. and R.Z.; supervision, C.W. and R.Z.; funding acquisition, C.W. All authors have read and agreed to the published version of the manuscript.

Funding: This research was funded by the Natural Scientific Foundation of Shandong (ZR2020ZD13, ZR2022JQ07), the National Natural Science Foundation of China (22376218), the Fundamental Research Funds for the Central Universities (22CX03033A), and the College Student Innovation Project (202303060CX).

Institutional Review Board Statement: Not applicable.

Informed Consent Statement: Not applicable.

Data Availability Statement: Data are contained within the article.

Conflicts of Interest: The authors declare no conflicts of interest.

erences

Kang, S.; Tanaka, T.; Narazaki, M.; Kishimoto, T. Targeting Interleukin-6 Signaling in Clinic. *Immunity* **2019**, *50*, 1007–1023. [CrossRef] [PubMed]
Cancelliere, R.; Di Tinno, A.; Di Lellis, A.M.; Contini, G.; Micheli, L.; Signori, E. Cost-effective and disposable label-free voltammetric immunosensor for sensitive detection of interleukin-6. *Biosens. Bioelectron.* **2022**, *213*, 114467. [CrossRef] [PubMed]
Interleukin-6 Receptor Antagonists in Critically Ill Patients with COVID-19. *N. Engl. J. Med.* **2021**, *384*, 1491–1502. [CrossRef] [PubMed]
Smolen, J.S.; Beaulieu, A.; Rubbert-Roth, A.; Ramos-Remus, C.; Rovensky, J.; Alecock, E.; Woodworth, T.; Alten, R. Effect of interleukin-6 receptor inhibition with tocilizumab in patients with rheumatoid arthritis (OPTION study): A double-blind, placebo-controlled, randomised trial. *Lancet* **2008**, *371*, 987–997. [CrossRef] [PubMed]
Kumari, N.; Dwarakanath, B.S.; Das, A.; Bhatt, A.N. Role of interleukin-6 in cancer progression and therapeutic resistance. *Tumor Biol.* **2016**, *37*, 11553–11572. [CrossRef] [PubMed]
Dou, C.; Wu, Z.; Chen, W.; Yan, H.; Li, D.; You, X.-Q.; Chen, Y.-S.; Zhou, C.; Chen, S.; Zhuang, P.; et al. Au-functionalized wrinkle graphene biosensor for ultrasensitive detection of Interleukin-6. *Carbon* **2024**, *216*, 118556. [CrossRef]
Russell, C.; Ward, A.C.; Vezza, V.; Hoskisson, P.; Alcorn, D.; Steenson, D.P.; Corrigan, D.K. Development of a needle shaped microelectrode for electrochemical detection of the sepsis biomarker interleukin-6 (IL-6) in real time. *Biosens. Bioelectron.* **2019**, *126*, 806–814. [CrossRef] [PubMed]
Wang, Y.-C.; Lin, S.-W.; Wang, I.J.; Yang, C.-Y.; Hong, C.; Sun, J.-R.; Feng, P.-H.; Lee, M.-H.; Shen, C.-F.; Lee, Y.-T.; et al. Interleukin-6 Test Strip Combined With a Spectrum-Based Optical Reader for Early Recognition of COVID-19 Patients With Risk of Respiratory Failure. *Front. Bioeng. Biotechnol.* **2022**, *10*, 796996. [CrossRef] [PubMed]
Majdinasab, M.; Lamy de la Chapelle, M.; Marty, J.L. Recent Progresses in Optical Biosensors for Interleukin 6 Detection. *Biosensors* **2023**, *13*, 898. [CrossRef]
Fan, G.-C.; Ren, X.-L.; Zhu, C.; Zhang, J.-R.; Zhu, J.-J. A new signal amplification strategy of photoelectrochemical immunoassay for highly sensitive interleukin-6 detection based on TiO2/CdS/CdSe dual co-sensitized structure. *Biosens. Bioelectron.* **2014**, *59*, 45–53. [CrossRef]
Shao, Z.-H.; Mo, H.-L.; Zhao, X.; Xie, F.; Zhao, G. Atomic-precise Pt2Cu4 cluster-based fluorescent sensor for rapid interleukin-6 detection. *Anal. Methods* **2023**, *15*, 3161–3164. [CrossRef] [PubMed]
Mak, W.C.; Beni, V.; Turner, A.P.F. Lateral-Flow Technology: From Visual to Instrumental. *TrAC-Trend. Anal. Chem.* **2016**, *79*, 297–305. [CrossRef]
Boehringer, H.R.; O'Farrell, B.J. Lateral Flow Assays in Infectious Disease Diagnosis. *Clin. Chem.* **2022**, *68*, 52–58. [CrossRef]
Sohrabi, H.; Majidi, M.R.; Khaki, P.; Jahanban-Esfahlan, A.; de la Guardia, M.; Mokhtarzadeh, A. State of the art: Lateral flow assays toward the point-of-care foodborne pathogenic bacteria detection in food samples. *Compr. Rev. Food Sci. Food Saf.* **2022**, *21*, 1868–1912. [CrossRef] [PubMed]
Hu, J.; Zhang, Z.L.; Wen, C.Y.; Tang, M.; Wu, L.L.; Liu, C.; Zhu, L.; Pang, D.W. Sensitive and Quantitative Detection of C-Reaction Protein Based on Immunofluorescent Nanospheres Coupled with Lateral Flow Test Strip. *Anal. Chem.* **2016**, *88*, 6577–6584. [CrossRef] [PubMed]
Omidfar, K.; Riahi, F.; Kashanian, S. Lateral Flow Assay: A Summary of Recent Progress for Improving Assay Performance. *Biosensors* **2023**, *13*, 837. [CrossRef]
Yin, X.; Liu, S.; Kukkar, D.; Wang, J.; Zhang, D.; Kim, K.-H. Performance enhancement of the lateral flow immunoassay by use of composite nanoparticles as signal labels. *TrAC-Trend. Anal. Chem.* **2024**, *170*, 117441. [CrossRef]

18. Gao, Z.; Ye, H.; Tang, D.; Tao, J.; Habibi, S.; Minerick, A.; Tang, D.; Xia, X. Platinum-Decorated Gold Nanoparticles with I Functionalities for Ultrasensitive Colorimetric in Vitro Diagnostics. *Nano Lett.* **2017**, *17*, 5572–5579. [CrossRef] [PubMed]
19. Wei, Z.; Luciano, K.; Xia, X. Catalytic Gold-Iridium Nanoparticles as Labels for Sensitive Colorimetric Lateral Flow Assay. *Nano* **2022**, *16*, 21609–21617. [CrossRef]
20. Rahbar, M.; Wu, Y.; Subramony, J.A.; Liu, G. Sensitive Colorimetric Detection of Interleukin-6 via Lateral Flow Assay Incorpor Silver Amplification Method. *Front. Bioeng. Biotechnol.* **2021**, *9*, 778269. [CrossRef]
21. Panferov, V.G.; Safenkova, I.V.; Varitsev, Y.A.; Drenova, N.V.; Kornev, K.P.; Zherdev, A.V.; Dzantiev, B.B. Development of Sensitive Lateral Flow Immunoassay with Silver Enhancement for the Detection of *Ralstonia solanacearum* in Potato Tubers. *Ta* **2016**, *152*, 521–530. [CrossRef] [PubMed]
22. Lei, R.; Arain, H.; Obaid, M.; Sabhnani, N.; Mohan, C. Ultra-Sensitive and Semi-Quantitative Vertical Flow Assay for the R Detection of Interleukin-6 in Inflammatory Diseases. *Biosensors* **2022**, *12*, 756. [CrossRef] [PubMed]
23. Lei, R.; Wang, D.; Arain, H.; Mohan, C. Design of Gold Nanoparticle Vertical Flow Assays for Point-of-Care Testing. *Diagno* **2022**, *12*, 1107. [CrossRef]
24. Jiang, N.; Ahmed, R.; Damayantharan, M.; Ünal, B.; Butt, H.; Yetisen, A.K. Lateral and Vertical Flow Assays for Point-of-C Diagnostics. *Adv. Healthc. Mater.* **2019**, *8*, 1900244. [CrossRef] [PubMed]
25. Han, Q.; Fan, L.; Liu, X.; Tang, Y.; Wang, P.; Shu, Z.; Zhang, W.; Zhu, L. Lateral Flow Immunoassay Based on Quantum Nanobeads for Detection of Chloramphenicol in Aquatic Products. *Molecules* **2023**, *28*, 7496. [CrossRef]
26. Liu, H.; Liu, Y.; Zhou, T.; Zhou, P.; Li, J.; Deng, A. Ultrasensitive and Specific Detection of Anticancer Drug 5-Fluorourac Blood Samples by a Surface-Enhanced Raman Scattering (SERS)-Based Lateral Flow Immunochromatographic Assay. *Mole* **2022**, *27*, 4019. [CrossRef]
27. Hu, J.; Jiang, Y.Z.; Tang, M.; Wu, L.L.; Xie, H.Y.; Zhang, Z.L.; Pang, D.W. Colorimetric-Fluorescent-Magnetic Nanosphere-B Multimodal Assay Platform for Salmonella Detection. *Anal. Chem.* **2019**, *91*, 1178–1184. [CrossRef]
28. Zhu, J.; Guo, G.; Liu, J.; Li, X.; Yang, X.; Liu, M.; Fu, C.; Zeng, J.; Li, J. One-pot synthesized Au@Pt nanostars-based lateral immunoassay for colorimetric and photothermal dual-mode detection of SARS-CoV-2 nucleocapsid antibody. *Anal. Chim.* **2024**, *1292*, 342241. [CrossRef]
29. Bradley, Z.; Coleman, P.A.; Courtney, M.A.; Fishlock, S.; McGrath, J.; Uniacke-Lowe, T.; Bhalla, N.; McLaughlin, J.A.; Hoga Hanrahan, J.P.; et al. Effect of Selenium Nanoparticle Size on IL-6 Detection Sensitivity in a Lateral Flow Device. *ACS Omega* *8*, 8407–8414. [CrossRef]
30. Zhou, W.; Hu, K.Q.; Kwee, S.; Tang, L.; Wang, Z.H.; Xia, J.F.; Li, X.J. Gold Nanoparticle Aggregation-Induced Quantita Photothermal Biosensing Using a Thermometer: A Simple and Universal Biosensing Platform. *Anal. Chem.* **2020**, *92*, 2739– [CrossRef]
31. Zhang, D.; Du, S.; Su, S.; Wang, Y.; Zhang, H. Rapid detection method and portable device based on the photothermal effe gold nanoparticles. *Biosens. Bioelectron.* **2019**, *123*, 19–24. [CrossRef]
32. Liang, J.; Wu, L.; Wang, Y.; Liang, W.; Hao, Y.; Tan, M.; He, G.; Lv, D.; Wang, Z.; Zeng, T.; et al. SERS/photothermal-b dual-modal lateral flow immunoassays for sensitive and simultaneous antigen detection of respiratory viral infections. *Actuators B Chem.* **2023**, *389*, 133875. [CrossRef]
33. Wang, K.; Liu, X.; Liang, X.; Jiang, Y.; Wen, C.-Y.; Zeng, J. Near-Infrared Responsive Ag@Au Nanoplates with Exceptional Stab for Highly Sensitive Colorimetric and Photothermal Dual-Mode Lateral Flow Immunoassay. *Anal. Chem.* **2024**, *96*, 3208– [CrossRef]
34. Wang, Z.; Wang, M.; Wang, X.; Hao, Z.; Han, S.; Wang, T.; Zhang, H. Photothermal-based nanomaterials and photothermal-sen An overview. *Biosens. Bioelectron.* **2023**, *220*, 114883. [CrossRef] [PubMed]
35. Xu, S.; Bai, X.; Wang, L. Exploration of photothermal sensors based on photothermally responsive materials: A brief review. *Chem. Front.* **2018**, *5*, 751–759. [CrossRef]
36. Zhang, Y.; Yang, H.-J.; Xu, Z.; Liu, X.; Zhou, J.; Qu, X.-F.; Wang, W.-L.; Feng, Y.; Peng, C. An ultra-sensitive photothermal la flow immunoassay for 17β-estradiol in food samples. *Food Chem.* **2023**, *404*, 134482. [CrossRef] [PubMed]
37. Zhang, Y.; Wang, Z.; Qu, X.; Zhou, J.; Yang, H.; Wang, W.; Yang, C. A photothermal lateral flow immunoassay for zearale with high sensitivity and wide detection range. *Sens. Actuators B Chem.* **2023**, *390*, 133909. [CrossRef]
38. Hao, W.; Huang, Y.; Wang, L.; Liang, J.; Yang, S.; Su, L.; Zhang, X. Smartphone-Based Photothermal Lateral Flow Immunoa Using Rhenium Diselenide Nanosheet. *ACS Appl. Mater. Interfaces* **2023**, *15*, 9665–9674. [CrossRef]
39. Wen, C.Y.; Zhao, L.J.; Wang, Y.; Wang, K.; Li, H.W.; Li, X.; Zi, M.; Zeng, J.B. Colorimetric and photothermal dual-mode lateral immunoassay based on Au-Fe_3O_4 multifunctional nanoparticles for detection of Salmonella typhimurium. *Microchim. Acta* *190*, 57. [CrossRef]
40. Guo, G.; Zhao, T.; Sun, R.; Song, M.; Liu, H.; Wang, S.; Li, J.; Zeng, J. Au-Fe_3O_4 dumbbell-like nanoparticles based lateral immunoassay for colorimetric and photothermal dual-mode detection of SARS-CoV-2 spike protein. *Chin. Chem. Lett.* *35*, 109198. [CrossRef]

Moreau, L.M.; Schurman, C.A.; Kewalramani, S.; Shahjamali, M.M.; Mirkin, C.A.; Bedzyk, M.J. How Ag Nanospheres Are Transformed into AgAu Nanocages. *J. Am. Chem. Soc.* **2017**, *139*, 12291–12298. [CrossRef] [PubMed]
Li, X.; Yu, D.; Li, H.; Sun, R.; Zhang, Z.; Zhao, T.; Guo, G.; Zeng, J.; Wen, C.-Y. High-density Au nanoshells assembled onto Fe_3O_4 nanoclusters for integrated enrichment and photothermal/colorimetric dual-mode detection of SARS-CoV-2 nucleocapsid protein. *Biosens. Bioelectron.* **2023**, *241*, 115688. [CrossRef] [PubMed]

molecules

Article

A Dopamine Detection Sensor Based on Au-Decorated NiS_2 a Its Medical Application

Chongchong Ma [1], Yixuan Wen [1], Yuqing Qiao [1], Kevin Z. Shen [2] and Hongwen Yuan [3,*]

1 State Key Laboratory of Metastable Materials Science and Technology, Yanshan University, Qinhuangdao 066004, China; macc321@163.com (C.M.); yixuanw001@163.com (Y.W.); qiaoyq@ysu.edu.cn (Y.Q.)
2 Department of Biology Texas, A&M University, College Station, TX 77483, USA; kevinxshen@gmail.com
3 School of Traditional Chinese Medicine, Capital Medical University, Beijing 100069, China
* Correspondence: yuanhw@ccmu.edu.cn

Abstract: This article reports a simple hydrothermal method for synthesizing nickel disulfide (NiS_2) on the surface of fluorine-doped tin oxide (FTO) glass, followed by the deposition of 5 nm Au nanoparticles on the electrode surface by physical vapor deposition. This process ensures the uniform distribution of Au nanoparticles on the NiS_2 surface to enhance its conductivity. Finally, an Au@NiS_2-FTO electrochemical biosensor is obtained for the detection of dopamine (DA). The composite material is characterized using transmission electron microscopy (TEM), UV-Vis spectroscopy, X-ray diffraction, and X-ray photoelectron spectroscopy. The electrochemical properties of the sensor are investigated using cyclic voltammetry (CV), differential pulse voltammetry (DPV), and time current curves in a 0.1 M PBS solution (pH = 7.3). In the detection of DA, Au@NiS_2-FTO exhibits a wide linear detection range (0.1~1000 μM), low detection limit (1 nM), and fast response time (0.1 s). After the addition of interfering substances, such as glucose, L-ascorbic acid, uric acid, $CaCl_2$, NaCl, and KCl, the electrode potential remains relatively unchanged, demonstrating its strong anti-interference capability. It also demonstrates strong sensitivity and reproducibility. The obtained Au@NiS_2-FTO provides a simple and easy-to-operate example for constructing nanometer catalysts with enzyme-like properties. These results provide a promising method utilizing Au coating to enhance the conductivity of transition metal sulfides.

Keywords: dopamine sensor; NiS_2; Au decorated; hydrothermal

1. Introduction

DA is a monoamine neurotransmitter found in the human brain and kidneys. It plays an important role in brain control, the immune system, and the renal system. It is crucial for human learning, cognition, motor behavior, and hormone production [1–4]. The DA level adjustment can lead to various neurological disorders, such as schizophrenia, Parkinson's disease, epilepsy, increased heart rate, arrhythmia, elevated blood pressure, and abnormalities in the cardiovascular system [5–8]. Therefore, determining and monitoring DA levels in real time is quite significant and can be used not only for early diagnosis of diseases but also for evaluating the effectiveness of various drugs in DA manipulation. Moreover, it helps to deepen our understanding of the mechanisms of DA function and better treatment of various diseases [9].

Traditional methods for detecting DA include blood tests and 24 h urine sampling. Usually, these methods require specialized technical training and laboratory equipment, including high-performance liquid chromatography, fluorescence spectroscopy, enzyme analysis, mass spectrometry, and capillary electrophoresis, which are greatly limited due to the high cost and operationally complex [10–14]. Recently, electrochemical sensing methods developed fast, offering advantages such as simplicity, low cost, ease of use, rapid analysis, and high sensitivity to compensate for the limitations of traditional methods. Currently, these sensors have been widely used in various fields such as biological analysis, clinical diagnosis, drug screening, and environmental monitoring. For instance, they can be utilized to detect biomolecules, like blood glucose, cholesterol, and tumor markers, enabling diabetes management, cardiovascular disease diagnosis, and early cancer detection. Moreover, electrochemical biosensors can also monitor heavy metal ions, organic pollutants, and microorganisms in the environment, contributing to environmental protection and food safety control. Ferlazzo et al. developed a yttria-stabilized zirconia electrochemical sensor for the precise determination of tyrosine [15]; Ahmed et al. prepared $In_2O_3 \cdot ZnO$@MC for dopamine detection with a detection limit of 0.024 μM [16]. These sensors have been widely applied in various electrocatalytic fields, such as supercapacitors and lithium-ion batteries [17,18]. Moreover, they can be integrated into miniaturized devices for use in medical experiments and procedures [19–21]. Thus, plenty of electrode materials are developed for the detection of DA. Electrode materials must be able to detect and differentiate other molecules, providing a distinct peak for DA. Meanwhile, they also need high anti-interference properties, enabling independent detection of DA amidst the presence of multiple substances without interference. Additionally, they should have a wide detection range to meet the medical research needs regarding DA levels [22–26].

Currently, transition metal dichalcogenides (TMDs) are receiving significant attention due to their distinctive structural features, morphologies, conductivity properties, and their role in facilitating electron transfer as narrow bandgap semiconductors [27–29]. Among the amount of TMDs, nickel disulfide (NiS_2) attracts much attention due to its high electrochemical activity and relatively low synthesis costs. NiS_2 is a good electrocatalytic material, but its poor conductivity largely limits its application [30]. Its intrinsic conductivity is modest, posing limitations on electron transfer within electrochemical biosensor systems [31,32]. Therefore, enhancing electrical conductivity is crucial for improving catalytic activity. Referring to previous catalytic materials, Au coating is a very effective technique. Kim et al. developed a method of Au nanoparticle modification of MoS_2, significantly improving the material's charge transfer [33]. Atta et al. doped Au into polyvinyl alcohol/carboxymethylcellulose, greatly enhancing the material's electrical conductivity [34]. However, this study presents a novel approach wherein nickel disulfide is directly synthesized via a one-step hydrothermal method. We coated Au on the surface of NiS_2 using physical vapor deposition in this study, significantly enhancing its electrical conductivity. A 5 nm Au film is deposited onto the surface using a physical vapor deposition (PVD) system, resulting in the formation of Au@NiS_2. Furthermore, through tests, such as CV, it was found that the electrocatalytic biosensing performance is greatly enhanced. The surface-bound Au particles serve to augment the conductivity of NiS_2. This composite system not only exhibits enhanced stability but also facilitates electron transfer subsequent to DA molecule signal recognition. Thus, the Au@NiS_2 composite material offers considerable potential for advancing the development of high-sensitivity electrochemical biosensors, representing a promising candidate for the next generation of materials in this field.

2. Results and Discussion

2.1. Electrochemical Characterization

In the experiment, we first compare the electrochemical behavior of Au@NiS_2-FTO, NiS_2-FTO, Au-FTO, and pure FTO, respectively. It has already been reported that the oxidation reduction reaction of DA involves a two-electron transfer process, converting DA to DA quinone [35,36]. The schematic of the electrochemical setup, as depicted in Figure 1a, illustrates the foundation for exploring the process of the electrochemical reaction process. This experiment was conducted employing a sophisticated three-electrode system integrated into the CHI660E electrochemical workstation. In this study, an FTO substrate (10 × 6 mm) was selected as the pivotal working electrode (WE). Complementing this, a platinum plate was designated as the auxiliary electrode (CE), while an Ag/AgCl electrode played the indispensable reference electrode (RE). Initially, DA was meticulously dissolved in 0.1 M PBS (pH = 7.3) to establish a standardized 1 mM DA solution. Subsequently, the distinct performances of the various electrodes were examined through the application of cyclic voltammetry. By scrutinizing the cyclic voltammetry outcomes, it became feasible to gauge the intrinsic electrocatalytic prowess of the DA sensing electrode, primarily by assessing the DA oxidation peak potential. This pivotal metric served as a cornerstone for evaluating the efficacy and sensitivity of the electrode in facilitating the electrochemical detection of DA.

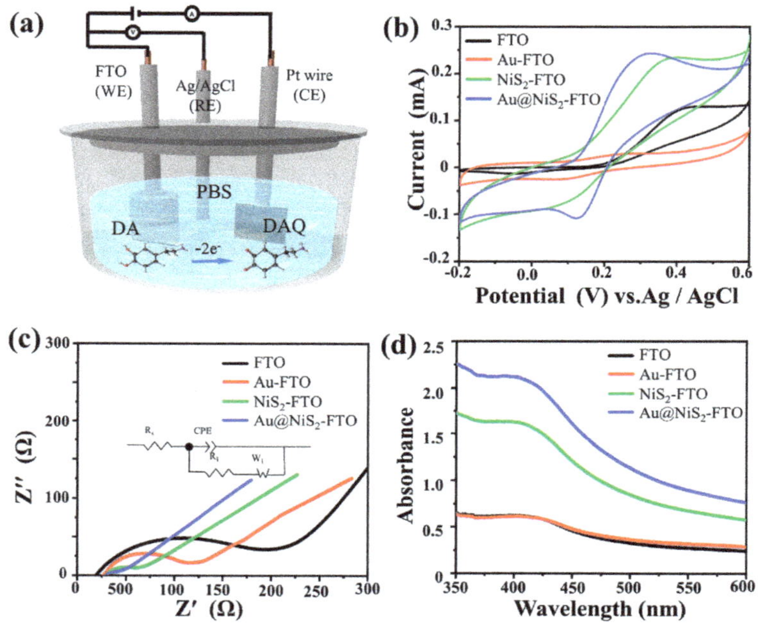

Figure 1. (**a**) Schematic of the electrochemical testing setup; (**b**) CVs of bare FTO, Au-FTO, NiS_2-FTO, and Au@NiS_2-FTO in 1 mM DA in 0.01 M PBS (pH = 7.3) at a scan rate of 0.1 V s^{-1}; (**c**) Nyquist plots of bare FTO, Au-FTO, FTO-NiS_2, and FTO-Au@NiS_2 in 0.1 M KCl containing 0.5 mM $Fe(CN)_6^{3-}$ and $Fe(CN)_6^{4-}$; (**d**) the UV–Vis absorption spectra of 1 mM DA in 0.1 M PBS (pH = 7.3) before and after applying a voltage of 0.3 V.

As shown in Figure 1b, the pristine NiS_2-FTO exhibits weak catalytic ability for the detection of DA, with a low peak and almost no reaction with DA observed for the original FTO and Au-FTO. In contrast, Au@NiS_2-FTO shows a significant current response to DA, with an oxidation potential reaching Epa = 0.3 V. The Au nanoparticles effectively enhance the electrocatalytic performance of NiS_2.

It is found that the Au coating significantly enhances the activity of NiS_2, and the large number of exposed active sites facilitates interaction with DA, thereby improving electrochemical performance [37,38]. As expected, the oxidation peak current of Au@NiS_2-FTO is 1.5 times higher than that of NiS_2-FTO, indicating that Au@NiS_2-FTO exhibits stronger electrocatalytic activity for DA detection. Compared to NiS_2-FTO, Au-FTO, and FTO, the peak separation (ΔEp) of Au@NiS_2-FTO is significantly reduced, suggesting improved reaction kinetics. Additionally, using a solution containing 5 mM $K_3[Fe(CN)_6]$, 5 mM $K_4[Fe(CN)_6]$, and 0.1 M KCl as the redox probe solution, the interface performance of different sensors was tested using electrochemical impedance spectroscopy (EIS). By fitting the EIS spectra with equivalent circuits (Figure 1c), the test data are shown in Table 1. The charge transfer resistance of the Au@NiS_2-FTO electrode is significantly lower than that of

NiS_2-FTO, Au-FTO, and FTO, with the sequence being FTO > Au-FTO > NiS_2-FTO. This is consistent with the CV results, indicating that Au@NiS_2-FTO has a better electron transfer rate and is, therefore, more suitable for DA detection. To further verify the oxidation reduction reaction between DA and DA quinone, we conducted tests using UV–visible absorption spectroscopy. In Figure 1b, it is observed that the oxidation peak of DA is at 0.3 V. Therefore, in this study, chronoamperometry was used to electrolyze 1 mM DA by different electrodes in three-electrode systems for 10 min, followed by testing the collected reaction solution, as shown in Figure 1d.

Table 1. The obtained R_{CT} values correspond to the Faradaic process taking place at various modified electrodes.

Electrode	R_{CT} (Ω)
FTO	153
Au-FTO	83
NiS_2-FTO	31
Au@NiS_2-FTO	20

In Figure 1d, it can be observed that FTO and Au-FTO did not exhibit significant absorption peaks. However, after electrolyzing for 10 min, a peak corresponding to the characteristic absorption peak of DA quinone was observed at around 400 nm for NiS_2-FTO and Au@NiS_2-FTO. Moreover, the absorption peak intensity of Au@NiS_2-FTO was significantly higher than that of NiS_2-FTO, indicating that the catalytic activity of Au@NiS_2-FTO is much higher than that of NiS_2-FTO. Additionally, no characteristic absorption peak of DA quinone was observed in the spectrum after electrolyzing for 10 min using Au-FTO, indicating that only Au@NiS_2-FTO exhibits catalytic activity.

2.2. Structure and Morphology Characterization

The electrocatalytic performance of materials is intricately intertwined with their underlying structure and morphology. In our experiment, we scrutinized the morphology and structure of the Au@NiS_2 material through transmission electron microscopy (TEM) and High-Resolution Transmission Electron Microscopy (HRTEM). Figure 2b vividly illustrates the characteristics of NiS_2, showcasing a remarkable two-dimensional nanosheet-like structure distribution, with Au nanofilm equably dispersed across the surface of NiS_2. Upon examination through HRTEM, the lattice spacing of 2.35 Å was linked to the (111) crystal plane of NiS_2, while the lattice spacing of 2.54 Å corresponded to the (210) crystal plane of Au. This inspection unequivocally demonstrated the seamless integration of Au with NiS_2, a pivotal aspect of catalytic efficacy. Further insights were gleaned from the Selected Area Electron Diffraction (SAED) pattern, revealing distinctive rings corresponding to the face-centered cubic structure reflections of Au, notably the (111), (220), and (311) planes, alongside the crystal planes of NiS_2, such as (121), (220), and (311). Additionally, the Energy-Dispersive X-ray Spectroscopy (EDX) spectra showcased the elemental distribution, and Au prominently adhered to the surface of NiS_2, showcasing a robust and well-integrated bond between the two materials. This comprehensive characterization not only elucidated the structure but also underscored the synergistic relationship between Au and NiS_2 in the context of electrocatalytic applications.

Figure 3a shows the X-ray diffraction (XRD) analysis of Au@NiS_2 into the structure of the composite material. Notably, the peaks corresponding to the presence of Au nanoparticles are discernible at 38.18° and 77.56°. By referencing the standard PDF card (PDF#99-0056), these peaks align precisely with the (111) and (311) crystal planes of Au, elucidating the crystalline orientation of the Au nanoparticles within the composite. Simultaneously, the XRD pattern also reveals characteristic peaks associated with NiS_2 at 27.25°, 31.59°, and 35.31°. These peaks are in direct correspondence with the (111), (200), and (210) crystal planes of NiS_2, as validated by the reference standard PDF card (PDF#11-0099).

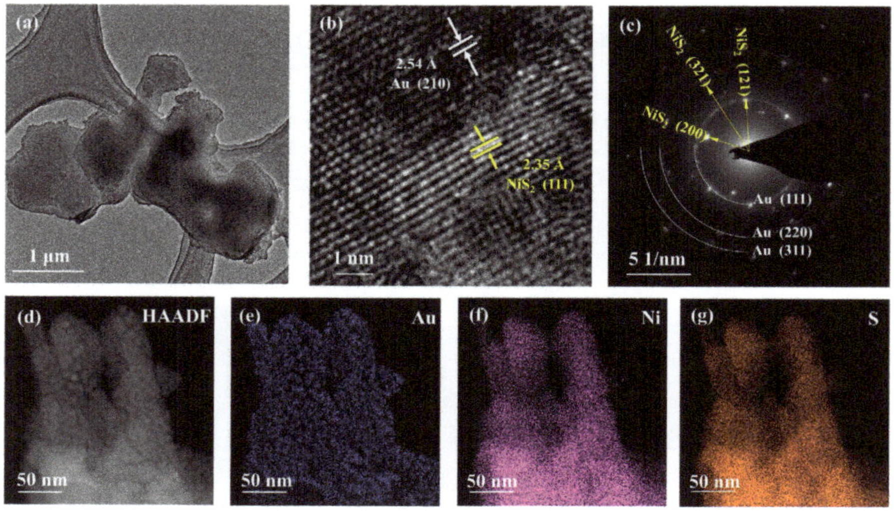

Figure 2. (**a**) TEM images of Au@NiS_2; (**b**) HRTEM images of Au@NiS_2; (**c**) SAED pattern of Au@NiS_2; (**d**) scanning HAADF-STEM image of Au@NiS_2 and corresponding elemental mapping of (**e**) Au, (**f**) Ni, and (**g**) S.

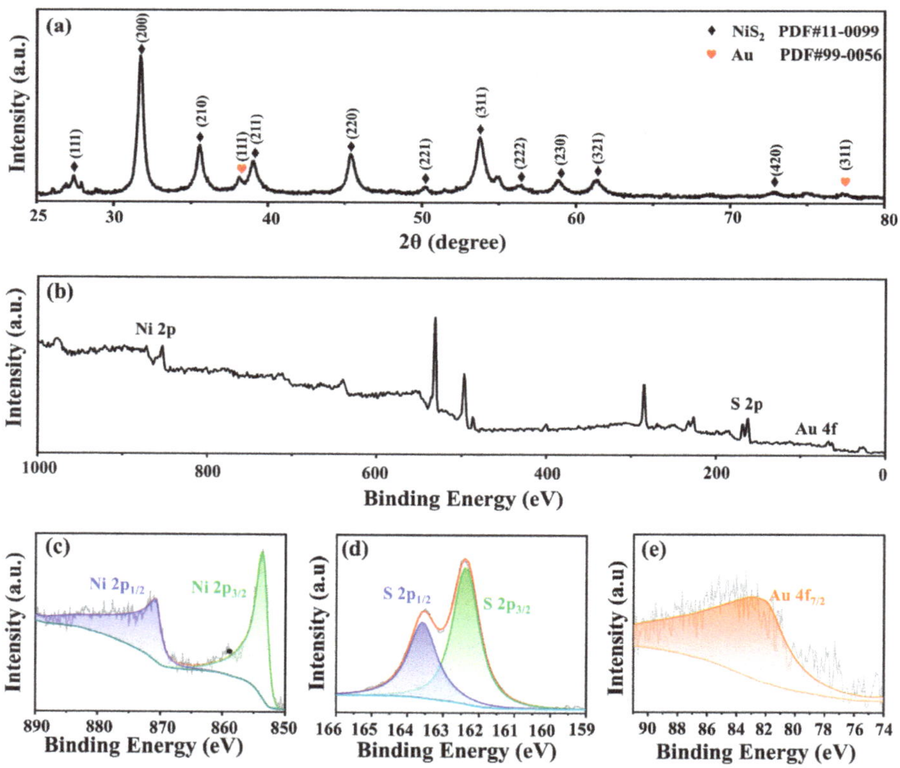

Figure 3. (**a**) XRD patterns of Au@NiS_2. (**b**) XPS survey spectra of Au@NiS_2 and the high-resolution spectra of Au@NiS_2 in the energy range of (**c**) Ni 2p, (**d**) S 2p, and (**e**) Au 4f.

The comprehensive characterization of the Au@NiS_2 composite material extended to the elemental composition and electronic states through X-ray photoelectron spectroscopy (XPS) analysis. Figure 3b shows the XPS spectrum of Au@NiS_2, unveiling distinctive peaks corresponding to the elemental constituents of Au, Ni, and S. Within the Au 4f spectrum, a peak emerges at 83.4 eV, indicative of the electronic configuration of Au within the composite. Transitioning to the Ni 2p spectrum, the presence of two distinct spin–orbit peaks, $2p_{3/2}$ and $2p_{1/2}$, is prominently observed at 855 and 872 eV. Furthermore, the corresponding peaks for these spin–orbit orbitals manifest at 162.5 and 163.8 eV, providing evidence that nickel predominantly exists in the Ni^{2+} oxidation state. This finding aligns with the structural revelations from the XRD analysis, thereby establishing a coherent link between the crystallographic characteristics and the electronic states of Ni within the Au@NiS_2 composite material [39].

2.3. Sensor Application

To delve deeper into the intricate dynamics of the electrochemical reaction kinetics between DA and Au@NiS_2-FTO, as depicted in Figure 4a, a meticulous exploration was undertaken, employing a three-electrode system immersed in a 0.1 M PBS (pH = 7.3) solution as the electrolyte medium. The experimental investigation entailed the application of cyclic voltammetry, encompassing a scan rate range spanning from 5 mV s^{-1} to 1 V s^{-1} and a voltage window extending from −0.2 to 0.6 V. The outcomes of these rigorous cyclic voltammetry experiments unveiled that the oxidation reduction peaks of DA exhibited a notable augmentation concomitant with the escalation of the scan rate. Simultaneously, a discernible shift in the oxidation potential of DA was observed, indicating a quasi-reversible nature of the electrochemical reaction within this system. These findings collectively underscore the dynamic interplay between DA and the Au@NiS_2-FTO composite electrode, shedding light on the kinetic intricacies governing the electrochemical processes at play. The anodic peak current (Ipa) and cathodic peak current (Ipc) are linearly related to the square root of the scan rate, and the linear regression equations are as follows:

$$\text{Ipa} = 0.022x - 0.04\ (R^2 = 0.99), \quad (1)$$

$$\text{Ipc} = -0.02x + 0.12\ (R^2 = 0.99), \quad (2)$$

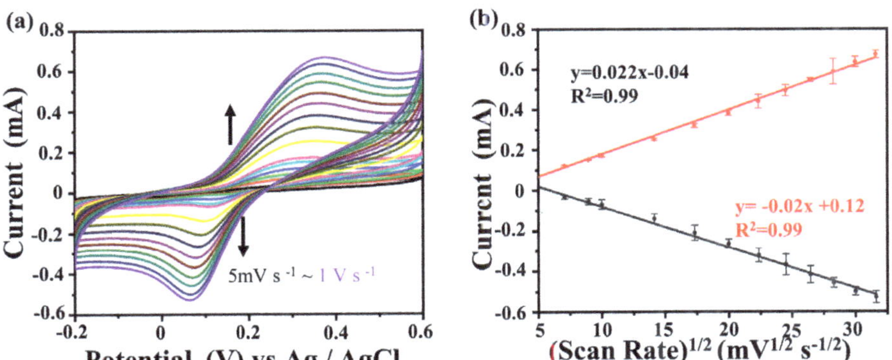

Figure 4. (**a**) CV curves of the Au@NiS_2-FTO-modified electrode in 0.1 M PBS (pH = 7.3) containing 1 mM DA by varying the scan rates from 5 mV s^{-1} to 1 V s^{-1}; (**b**) Ipa and Ipc vs. $v^{1/2}$.

The slope of this regression curve indicates that this electrochemical behavior is mainly controlled by diffusion, and its reversibility is good [40–42]. In the oxidation reaction, DA is converted to DA quinone, which is then reduced back to DA in the electrochemical reaction [43].

Subsequently, in order to further determine the response of Au@NiS_2-FTO to different concentrations of DA solutions, a standard three-electrode system was used with 0.1 M PBS (pH = 7.3) as the electrolyte. Cyclic voltammetry was performed within the voltage range of −0.2 V to 0.6 V at the same scan rate (0.1 V s^{-1}) to test the response signal to the addition of different concentrations of DA, as shown in Figure 5a. The linear regression equations obtained from cyclic voltammetry are as follows:

$$\text{Ipa} = 0.00025\text{x} + 0.0438\ (R^2 = 0.99), \tag{3}$$

$$\text{Ipc} = -0.00011\text{x} - 0.03718\ (R^2 = 0.99), \tag{4}$$

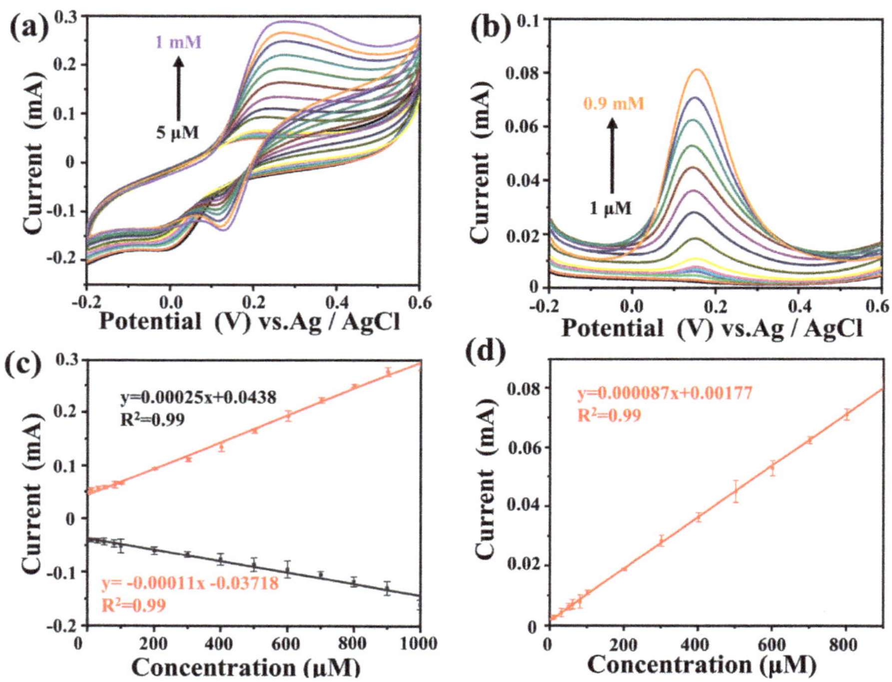

Figure 5. (**a**) CV curves of Au@NiS_2-FTO in 0.1 M PBS (pH = 7.3) containing various concentrations of DA (5–1 mM) at a scan rate of 0.1 V s^{-1}; (**b**) DPV curves of Au@NiS_2-FTO in 0.1 M PBS (pH = 7.3) containing various concentrations of DA (1–0.9 mM); (**c**) linear relationship between peak potential and concentration in CV; (**d**) linear relationship between peak potential and concentration in DPV.

In (3) and (4), it can be seen that Au@NiS_2-FTO exhibits a good response signal to DA, and under the stimulation of different concentrations of DA, the oxidation reduction peak remains unchanged, indicating that Au@NiS_2-FTO owns a stable ability for the detection of DA. The DA concentration range satisfied by this regression curve is 5 μM to 1 mM, indicating that Au@NiS_2-FTO can achieve broad-range DA detection.

Moreover, differential pulse voltammetry (DPV) is an electrochemical analysis technique commonly used to measure electrochemically active substances in solution. The testing method involves applying a series of pulse voltages on an electrochemical electrode and measuring the current response caused by these pulse voltages. DPV is a widely used electrochemical measurement method in sensor analysis, known for its high sensitivity and selectivity [44]. DPV was used to test its sensitivity by adding different concentrations of DA within the same voltage range. Similarly, in a three-electrode system with 0.1 M PBS (pH = 7.3) as the electrolyte, DPV was performed with a voltage window from −0.2 V to

0.6 V, a step potential of 4 mV, an amplitude of 50 mV, and a pulse period of 0.2 s. The obtained data are shown in Figure 5d, and the linear regression equation is as follows:

$$Ipc = -0.000087x + 0.00177\ (R^2 = 0.99), \quad (5)$$

In Figure 5b, it can be seen that the linear detection range of DPV is from 1 μM to 0.9 mM.

Figure 5 shows that the voltage peak value of DA conversion to DA quinone on the Au@NiS_2-FTO electrode is 0.3 V. Therefore, we use 0.3 V as the working potential; the time current curve of DA on the Au@NiS_2-FTO electrode was tested, with a testing range from 0.1 μM to 1 mM. The results are shown in Figure 6a. Regression curves of the current at different concentrations and at the same time (100 s) are plotted, as shown in Figure 6a. The linear regression equation is as follows:

$$I = 0.0000342x + 0.00153\ (R^2 = 0.99), \quad (6)$$

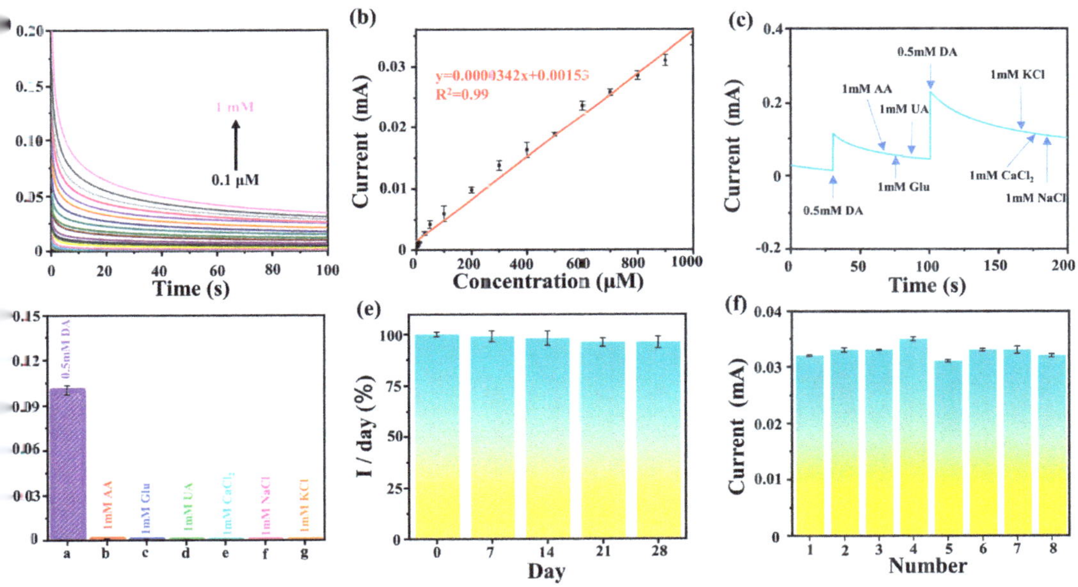

Figure 6. (**a**) The curves of Au@NiS_2-FTO in 0.1 M PBS (pH = 7.3) containing various concentrations of DA (0.1–1 mM) at 0.3 V; (**b**) linear relationship between peak potential and concentration in CV; (**c**) amperometric response obtained at Au@NiS_2-FTO for the successive addition of 0.5 mM DA, 1 mM glucose, 1 mM AA, 1 mM UA, and 0.5 mM DA into 0.1 mM PBS (pH = 7.3); (**d**) current density response columnar diagram of the tested analytes compared with DA; (**e**) stability of Au@NiS_2-FTO for DA detection; (**f**) reproducibility of Au@NiS_2-FTO biosensors.

In the calibration curve, it can be known that the concentration of DA is directly proportional to the absolute value of the oxidation current, and its linear detection range is from 0.1 μM to 1 mM and has a low detection limit (1 nM), indicating high sensitivity of the Au@NiS_2-FTO electrode for DA detection.

Another key aspect of biosensing technology is the selectivity for common interfering substances. In the extracellular fluid and the serum of the mammalian central nervous system, there are also some reducing agents with oxidation potentials close to that of DA, such as glucose, L-ascorbic acid, uric acid, Ca^{2+}, Na^+, and K^+, which may have a certain impact on the accurate determination of DA content. It is necessary to distinguish DA from these interfering substances and exploring the anti-interference ability of the Au@NiS_2-FTO electrode is extremely important [45–48].

Similarly, we employed a three-electrode system to investigate the interference of glucose, L-ascorbic acid, uric acid, $CaCl_2$, NaCl, and KCl on DA. The results are depicted in Figure 6c. Upon adding 0.5 mM DA to the electrolyte, an instantaneous change in the current occurred. Subsequently, the addition of 1 mM glucose, 1 mM L-ascorbic acid, 1 mM uric acid, 1 mM $CaCl_2$, 1 mM NaCl, and 1 mM KCl did not result in significant changes in the curve trend. However, upon further addition of 0.5 mM DA, the current exhibited significant fluctuations, indicating that glucose, L-ascorbic acid, uric acid, $CaCl_2$, NaCl, and KCl did not induce changes in the current in this environment. To better observe the impact of adding interfering substances on DA, the current change at 0.2 s after adding interfering substances was plotted, as shown in Figure 6d. It is evident in the figure that the current variation in the system after adding interfering substances can be negligible, demonstrating the specificity and sensitivity of the Au@NiS_2-FTO electrode for DA detection. To test the stability of the Au@NiS_2-FTO electrode, we conducted tests on five sets of electrodes every 7 days to determine the response signal. The results, as shown in Figure 6e, indicate that the Au@NiS_2-FTO electrode can maintain over 96% of its activity after 28 days, demonstrating its stability in air. Subsequently, using eight independent electrodes under the same experimental parameters, we measured 1 mM DA, with the results shown in Figure 6f. Calculations show a relative standard deviation (RSD) of 3%, indicating good reproducibility of the biosensor. In conclusion, the proposed Au@NiS_2-FTO electrode exhibits excellent characteristics, such as anti-interference, stability, and reproducibility, proving its potential for measuring DA in real samples and paving the way for future in vivo testing of Au@NiS_2-FTO electrodes.

Table 2 summarizes the electrochemical detection performance of various biosensing materials. Compared with other sensing materials, the Au@NiS_2-FTO prepared in this study demonstrates a much wider detection range and lower limit of detection (LOD).

Table 2. A comparison of the analytical performance of the proposed sensor with other DA sensors.

Materials	Linear Range (μM)	LOD (nM)	Reference
Au@NiS_2-FTO	0.1–1000	1	This work
SWCNTs-GCE	0.5–100	190	[49]
ZnO@Au	0.1–500	8.5	[50]
$K_2Fe_4O_7$/GCE	1–140	22	[51]
Ag/CuO PNBs	0.04–10	7	[52]
Split aptamer sensor	5–50	1000	[53]
pS-BIL MIP PeGE	0.05–250	20	[44]
Gallic acid-RGO/AuNPs	0.01–100.3	2.6	[54]
WO_3 NPs-GCE	0.1–50, 50–600	24	[55]
CuO	5–40	110	[56]

3. Materials and Methods

3.1. Material Preparations

$NiCl_2·6H_2O$, $CaCl_2$, and NaCl were purchased from Inokai; $Na_2S_2O_3·5H_2O$, uric acid (UA), ascorbic acid (AA), the phosphate buffer solution (PBS, pH = 7.3) were purchased from Macklin; and KCl, $[K_3Fe(CN)_6]$, $[K_4Fe(CN)_6]·3H_2O$, and DA were purchased from Aladdin. All reagents used in the experiments were at an analytical grade level and required no further purification. Deionized water was used throughout the whole experiment.

In the experiment, NiS_2 was first prepared on the FTO substrate using a traditional hydrothermal method. $NiCl_2·6H_2O$ (1.188 g, 0.005 mol) and $Na_2S_2O_3·5H_2O$ (2.482 g, 0.010 mol) were placed into a 50 mL capacity polytetrafluoroethylene-lined autoclave. Then, 30 mL of ultrapure water was added and stirred for 30 min until dissolved to form a transparent green solution. FTO conductive glass was placed below the liquid level and fixed to the inner wall of the autoclave using heat-resistant tape. The autoclave was maintained at 150 °C for 12 h. Cool to room temperature and take out the sample. The as-prepared NiS_2 deposited on the FTO was washed repeatedly with distilled water and dried at 65 °C for 12 h.

3.2. Preparation of Au@NiS_2-FTO

A ZHDS400 high-vacuum organic/metal evaporation coating machine (PVD) was utilized for the preparation of Au nanofilm. Place the Au target in the crucible and place NiS_2-FTO on the sample plate using high-temperature-resistant tape and put it in the evaporation chamber. Heat the sample plate to 80 °C in a vacuum environment and adjust the power to 30 W. Evaporate for 2 min at a rate of 0.42 Å s^{-1}, depositing 5 nm of Au on the surface of NiS_2-FTO. Then, cool the sample to room temperature, deposit 5 nm of Au on the NiS_2-FTO surface, and obtain Au@NiS_2-FTO. As a control experiment, 5 nm Au is deposited on the FTO without modification of NiS_2 using the same method on the coating machine. In this experiment, we chose Au target material with a purity of 99.99%.

3.3. Electrochemical Performance Measurement

All the electrochemical tests mentioned in this article were conducted using the CHI660E electrochemical workstation produced by Shanghai Chenhua Instrument Co., Ltd., Shanghai, China. The platinum electrode and Ag/AgCl electrode were produced by Wuhan Gaoshiruilian Technology Co., Ltd., Wuhan, China, and the conductive glass FTO was purchased from Suzhou Shangyang Solar Technology Co., Ltd., Suzhou, China.

Electrochemical tests were conducted using a three-electrode system on a CHI 660E electrochemical workstation. An FTO substrate (10 × 6 mm) coated with NiS_2 was used as the working electrode (WE), a platinum plate served as the auxiliary electrode (CE), and an Ag/AgCl electrode was used as the reference electrode (RE). Differential pulse voltammetry (DPV) was performed in a 0.1 M PBS (pH = 7.3) solution as the supporting electrolyte.

In the experiment, the Au@NiS_2-FTO samples prepared by the hydrothermal method and magnetron sputtering deposition were used as working electrodes (WEs). Electrochemical tests, including CV and DPV, were carried out in a 0.1 M PBS (pH = 7.3) solution as the electrolyte, with a scan rate of 100 mV s^{-1} and a potential range of −0.2 to 0.6 V. EIS was carried out in a solution containing 5 mM [$K_3Fe(CN)_6$], 5 mM [$K_4Fe(CN)_6$], and 0.1 M KCl, with a frequency range of 0.01 to 100,000 Hz. Time current curve testing was performed with a voltage input of 0.3 V over a range of 100 s.

4. Conclusions

In summary, an Au@NiS_2-FTO was synthesized on the FTO substrate by a combining hydrothermal and PVD method. The structure and morphology of the samples were characterized by XRD, STEM, TEM, and XPS, which integrally confirmed that Au nanoparticles were uniformly distributed on the surface. The NiS_2 nanosheets own a large specific surface area and display a significant improvement in electrocatalytic activity towards DA. Moreover, the Au@NiS_2-FTO electrode showed excellent sensitivity (0.1~1000 μM) and a detection limit of 1 nM for DA, with an electrochemical response time of 0.1 s and outstanding stability. Even after the addition of interfering substances, such as glucose, L-ascorbic acid, uric acid, $CaCl_2$, NaCl, and KCl, its potential remained largely unaffected. After 28 days, the Au@NiS_2-FTO electrode still retains 96% activity, with high reproducibility and an RSD of 3%, demonstrating high resistance to interference, reproducibility, and stability. This work confirms that the Au surface modification is an effective approach to enhancing the sensing ability of DA. This provides direction for DA detection in real serum samples in the future.

Author Contributions: H.Y. conceived and designed the project. C.M. and Y.W. fabricated the samples. C.M. and Y.Q. carried out the biosensing experiments. K.Z.S. and H.Y. supervised the experiments. C.M. and H.Y. co-wrote the paper. All the authors discussed the results and commented on the manuscript. All authors have read and agreed to the published version of the manuscript.

Funding: This work was financially supported by the National Natural Science Foundation of China (No. 22279111).

Institutional Review Board Statement: Not applicable.

Informed Consent Statement: Not applicable.

Data Availability Statement: Data are contained within the article.

Conflicts of Interest: The authors declare no conflicts of interest.

References

1. Wang, Y.; Zhang, X.; Chen, Y.; Xu, H.; Tan, Y.; Wang, S. Detection of Dopamine Based on Tyrosinase-Fe_3O_4 Nanoparticles-chit Nanocomposite Biosensor. *Am. J. Biomed. Sci.* **2010**, *2*, 209–216. [CrossRef]
2. Luo, Q.; Su, Y.; Zhang, H. Sensitive dopamine sensor based on electrodeposited gold nanoparticles and electro-modulated M nanoflakes. *J. Iran. Chem. Soc.* **2022**, *20*, 731–738. [CrossRef]
3. Jing, W.-J.; Li, F.-F.; Liu, Y.; Ma, R.-N.; Zhang, W.; Shang, L.; Li, X.-J.; Xue, Q.-W.; Wang, H.-S.; Jia, L.-P. An electrochem ratiometric biosensor for the detection of dopamine based on an MXene-Au nanocomposite. *Chem. Commun.* **2023**, *59*, 12911–1 [CrossRef] [PubMed]
4. Yu, D.; Zhang, F.; Zhang, Y.; Lin, H.; Guo, W.; Yu, K.; Qu, F. Heterophase-Structured Cobalt Hydroxide on Partly Red Graphene Oxide for Enhanced Dopamine Biosensing. *ACS Appl. Eng. Mater.* **2023**, *1*, 1963–1972. [CrossRef]
5. Gong, W.; Li, J.; Chu, Z.; Yang, D.; Subhan, S.; Li, J.; Huang, M.; Zhang, H.; Zhao, Z. A low-cost high-entropy porous CrO/Cr biosensor for highly sensitive simultaneous detection of dopamine and uric acid. *Microchem. J.* **2022**, *175*, 107188. [CrossRe
6. Xu, C.; Gu, C.; Xiao, Q.; Chen, J.; Yin, Z.-Z.; Liu, H.; Fan, K.; Li, L. A highly selective and sensitive biosensor for dopamine b on a surface molecularly imprinted layer to coordinate nano-interface functionalized acupuncture needle. *Chem. Eng. J.* **2022**, 135203. [CrossRef]
7. Sun, Z.; Sun, S.; Jiang, X.; Ai, Y.; Xu, W.; Xie, L.; Sun, H.-B.; Liang, Q. Oligo-layer graphene stabilized fully exposed Fe-site ultra-sensitivity electrochemical detection of dopamine. *Biosens. Bioelectron.* **2022**, *211*, 114367. [CrossRef] [PubMed]
8. Yue, H.Y.; Zhang, H.J.; Huang, S.; Lu, X.X.; Gao, X.; Song, S.S.; Wang, Z.; Wang, W.Q.; Guan, E.H. Highly sensitive and sele dopamine biosensor using Au nanoparticles-ZnO nanocone arrays/graphene foam electrode. *Mater. Sci. Eng. C* **2020**, *108*, 11 [CrossRef] [PubMed]
9. Jiang, Y.; Wang, B.; Meng, F.; Cheng, Y.; Zhu, C. Microwave-assisted preparation of N-doped carbon dots as a biosenso electrochemical dopamine detection. *J. Colloid Interface Sci.* **2015**, *452*, 199–202. [CrossRef]
10. Liu, X.; Hou, Y.; Chen, S.; Liu, J. Controlling dopamine binding by the new aptamer for a FRET-based biosensor. *Bio Bioelectron.* **2021**, *173*, 112798. [CrossRef]
11. Lakard, S.; Pavel, I.-A.; Lakard, B. Electrochemical Biosensing of Dopamine Neurotransmitter: A Review. *Biosensors* **2021**, *11* [CrossRef] [PubMed]
12. Akbar, F.; Kolahdouz, M.; Larimian, S.; Radfar, B.; Radamson, H.H. Graphene synthesis, characterization and its applicatio nanophotonics, nanoelectronics, and nanosensing. *J. Mater. Sci. Mater. Electron.* **2015**, *26*, 4347–4379. [CrossRef]
13. Wang, K.; Liu, P.; Ye, Y.; Li, J.; Zhao, W.; Huang, X. Fabrication of a novel laccase biosensor based on silica nanoparticles mod with phytic acid for sensitive detection of dopamine. *Sens. Actuators B Chem.* **2014**, *197*, 292–299. [CrossRef]
14. Park, S.J.; Song, H.S.; Kwon, O.S.; Chung, J.H.; Lee, S.H.; An, J.H.; Ahn, S.R.; Lee, J.E.; Yoon, H.; Park, T.H.; et al. Hu dopamine receptor nanovesicles for gate-potential modulators in high-performance field-effect transistor biosensors. *Sci.* **2014**, *4*, 4342. [CrossRef] [PubMed]
15. Ferlazzo, A.; Espro, C.; Iannazzo, D.; Bonavita, A.; Neri, G. Yttria-zirconia electrochemical sensor for the detection of tyro *Mater. Today Commun.* **2023**, *35*, 106036. [CrossRef]
16. Ahmed, J.; Faisal, M.; Algethami, J.S.; Alsaiari, M.; Harraz, F.A. A novel In_2O_3-doped ZnO decorated mesoporous ca nanocomposite as a sensitive and selective dopamine electrochemical sensor. *J. Mater. Res. Technol.* **2024**, *29*, 540–549. [Cros
17. Wang, S.; Ning, P.; Huang, S.; Wang, W.; Fei, S.; He, Q.; Zai, J.; Jiang, Y.; Hu, Z.; Qian, X.; et al. Multi-functional NiS_2/FeS doped carbon nanorods derived from metal-organic frameworks with fast reaction kinetics for high performance overall w splitting and lithium-ion batteries. *J. Power Sources* **2019**, *436*, 226857. [CrossRef]
18. Wang, Y.; Cai, Z.; Duan, H.; Zhang, F.; Zhai, B.; Zhao, J.; Wang, X. Controlled synthesis of rod-like three-dimensi NiS_2/graphene nanostructures from metal complexes and their application in supercapacitor electrodes. *J. Phys. Chem. S* **2022**, *167*, 110716. [CrossRef]
19. Vellaichamy, B.; Periakaruppan, P.; Paulmony, T. Evaluation of a New Biosensor Based on in Situ Synthesized PPy-Ag- Nanohybrid for Selective Detection of Dopamine. *J. Phys. Chem. B* **2017**, *121*, 1118–1127. [CrossRef]
20. Kajisa, T.; Li, W.; Michinobu, T.; Sakata, T. Well-designed dopamine-imprinted polymer interface for selective and quantit dopamine detection among catecholamines using a potentiometric biosensor. *Biosens. Bioelectron.* **2018**, *117*, 810–817. [Cros
21. Shin, J.-W.; Yoon, J.; Shin, M.; Choi, J.-W. Electrochemical Dopamine Biosensor Composed of Silver Encapsulated MoS_2 Hy Nanoparticle. *Biotechnol. Bioprocess Eng.* **2019**, *24*, 135–144. [CrossRef]

Dong, X.; Lu, X.; Zhang, K.; Zhang, Y. Chronocoulometric DNA biosensor based on a glassy carbon electrode modified with gold nanoparticles, poly(dopamine) and carbon nanotubes. *Microchim. Acta* **2012**, *180*, 101–108. [CrossRef]

Li, M.; Liu, C.; Zhao, H.; An, H.; Cao, H.; Zhang, Y.; Fan, Z. Tuning sulfur doping in graphene for highly sensitive dopamine biosensors. *Carbon* **2015**, *86*, 197–206. [CrossRef]

Ghadimi, H.; Mahmoudian, M.R.; Basirun, W.J. A sensitive dopamine biosensor based on ultra-thin polypyrrole nanosheets decorated with Pt nanoparticles. *RSC Adv.* **2015**, *5*, 39366–39374. [CrossRef]

Arya Nair, J.S.; Saisree, S.; Aswathi, R.; Sandhya, K.Y. Ultra-selective and real-time detection of dopamine using molybdenum disulphide decorated graphene-based electrochemical biosensor. *Sens. Actuators B Chem.* **2022**, *354*, 131254. [CrossRef]

Yuan, Y.; Wang, S.; Wu, P.; Yuan, T.; Wang, X. Lignosulfonate in situ-modified reduced graphene oxide biosensors for the electrochemical detection of dopamine. *RSC Adv.* **2022**, *12*, 31083–31090. [CrossRef] [PubMed]

Kang, T.; Tang, T.W.; Pan, B.; Liu, H.; Zhang, K.; Luo, Z. Strategies for Controlled Growth of Transition Metal Dichalcogenides by Chemical Vapor Deposition for Integrated Electronics. *ACS Mater. Au* **2022**, *2*, 665–685. [CrossRef] [PubMed]

Mondal, A.; Vomiero, A. 2D Transition Metal Dichalcogenides-Based Electrocatalysts for Hydrogen Evolution Reaction. *Adv. Funct. Mater.* **2022**, *32*, 2208994. [CrossRef]

Lu, T.; Wang, Y.; Cai, G.; Jia, H.; Liu, X.; Zhang, C.; Meng, S.; Liu, M. Synthesizability of transition-metal dichalcogenides: A systematic first-principles evaluation. *Mater. Futures* **2023**, *2*, 015001 [CrossRef]

Chen, S.; Pan, Y. Enhancing catalytic properties of noble metal@MoS_2/WS_2 heterojunction for the hydrogen evolution reaction. *Appl. Surf. Sci.* **2022**, *591*, 153168. [CrossRef]

Wei, C.; Cheng, C.; Cheng, Y.; Wang, Y.; Xu, Y.; Du, W.; Pang, H. Comparison of NiS_2 and α-NiS hollow spheres for supercapacitors, non-enzymatic glucose sensors and water treatment. *Dalton Trans.* **2015**, *44*, 17278–17285. [CrossRef] [PubMed]

Lu, Z.; Li, Y.; Liu, T.; Wang, G.; Sun, M.; Jiang, Y.; He, H.; Wang, Y.; Zou, P.; Wang, X.; et al. A dual-template imprinted polymer electrochemical sensor based on AuNPs and nitrogen-doped graphene oxide quantum dots coated on NiS_2/biomass carbon for simultaneous determination of dopamine and chlorpromazine. *Chem. Eng. J.* **2020**, *389*, 124417. [CrossRef]

Kim, J.; Byun, S.; Smith, A.J.; Yu, J.; Huang, J. Enhanced Electrocatalytic Properties of Transition-Metal Dichalcogenides Sheets by Spontaneous Gold Nanoparticle Decoration. *J. Phys. Chem. Lett.* **2013**, *4*, 1227–1232. [CrossRef] [PubMed]

Atta, M.R.; Alsulami, Q.A.; Asnag, G.M.; Rajeh, A. Enhanced optical, morphological, dielectric, and conductivity properties of gold nanoparticles doped with PVA/CMC blend as an application in organoelectronic devices. *J. Mater. Sci. Mater. Electron.* **2021**, *32*, 10443–10457. [CrossRef]

Chen, J.-L.; Yan, X.-P.; Meng, K.; Wang, S.-F. Graphene Oxide Based Photoinduced Charge Transfer Label-Free Near-Infrared Fluorescent Biosensor for Dopamine. *Anal. Chem.* **2011**, *83*, 8787–8793. [CrossRef] [PubMed]

Renganathan, V.; Balaji, R.; Chen, S.-M.; Singh, V. The electrochemical determination of hazardous 4-hydroxynitrobenzene using NiS2 decorated graphene oxide nanocomposite in the river water sample. *Microchem. J.* **2020**, *153*, 104502. [CrossRef]

Xia, N.; Deng, D.; Zhang, L.; Yuan, B.; Jing, M.; Du, J.; Liu, L. Sandwich-type electrochemical biosensor for glycoproteins detection based on dual-amplification of boronic acid-gold nanoparticles and dopamine-gold nanoparticles. *Biosens. Bioelectron.* **2013**, *43*, 155–159. [CrossRef] [PubMed]

Silva, T.R.; Vieira, I.C. A biosensor based on gold nanoparticles stabilized in poly(allylamine hydrochloride) and decorated with laccase for determination of dopamine. *Analyst* **2016**, *141*, 216–224. [CrossRef] [PubMed]

Kumar, D.R.; Baynosa, M.L.; Dhakal, G.; Shim, J.-J. Sphere-like Ni_3S_4/NiS_2/MoO_x composite modified glassy carbon electrode for the electrocatalytic determination of d-penicillamine. *J. Mol. Liq.* **2020**, *301*, 112447. [CrossRef]

Martín, M.; Salazar, P.; Villalonga, R.; Campuzano, S.; Pingarrón, J.M.; González-Mora, J.L. Preparation of core–shell Fe_3O_4@poly(dopamine) magnetic nanoparticles for biosensor construction. *J. Mater. Chem. B* **2014**, *2*, 739–746. [CrossRef]

Baloach, Q.-U.; Nafady, A.; Tahira, A.; Sirajuddin; Sherazi, S.T.H.; Shaikh, T.; Arain, M.; Willander, M.; Ibupoto, Z.H. An amperometric sensitive dopamine biosensor based on novel copper oxide nanostructures. *Microsyst. Technol.* **2016**, *23*, 1229–1235. [CrossRef]

Liu, Y.; Zhang, Y.; Wang, C.; Zeng, X.; Lei, J.; Hou, J.; Huo, D.; Hou, C. Co Single-Atom Nanozymes for the Simultaneous Electrochemical Detection of Uric Acid and Dopamine in Biofluids. *ACS Appl. Nano Mater.* **2024**, *7*, 6273–6283. [CrossRef]

Xie, Z.; Shao, M.; Liu, Z.; Ren, X.; Gao, M.; Ma, H.; Zhang, N.; Wei, Q. Ultrasensitive aggregation-induced electrochemiluminescence sensor for dopamine detection in polymer hydrogel system. *Sens. Actuators B Chem.* **2024**, *398*, 134781. [CrossRef]

Kaya, H.K.; Cinar, S.; Altundal, G.; Bayramlı, Y.; Unaleroglu, C.; Kuralay, F. A novel design thia-bilane structure-based molecular imprinted electrochemical sensor for sensitive and selective dopamine determination. *Sens. Actuators B Chem.* **2021**, *346*, 130425. [CrossRef]

Li, S.-M.; Wang, Y.-S.; Hsiao, S.-T.; Liao, W.-H.; Lin, C.-W.; Yang, S.-Y.; Tien, H.-W.; Ma, C.-C.M.; Hu, C.-C. Fabrication of a silver nanowire-reduced graphene oxide-based electrochemical biosensor and its enhanced sensitivity in the simultaneous determination of ascorbic acid, dopamine, and uric acid. *J. Mater. Chem. C* **2015**, *3*, 9444–9453. [CrossRef]

Rahman, S.F.; Min, K.; Park, S.-H.; Park, J.-H.; Yoo, J.C.; Park, D.-H. Highly sensitive and selective dopamine detection by an amperometric biosensor based on tyrosinase/MWNT/GCE. *Korean J. Chem. Eng.* **2016**, *33*, 3442–3447. [CrossRef]

Ghosh, D.; Tabassum, R.; Sarkar, P.P.; Rahman, M.D.A.; Jalal, A.H.; Islam, N.; Ashraf, A. Graphene Nanocomposite Ink Coated Laser Transformed Flexible Electrodes for Selective Dopamine Detection and Immunosensing. *ACS Appl. Bio Mater.* **2024**, *7*, 3143–3153. [CrossRef] [PubMed]

48. Karim, A.; Yasser, M.; Ahmad, A.; Natsir, H.; Wahid Wahab, A.; Fauziah, S.; Taba, P.; Pratama, I.; Rosalin; Rajab, A.; et review: Progress and trend advantage of dopamine electrochemical sensor. *J. Electroanal. Chem.* **2024**, *959*, 118157. [CrossR
49. Yang, J.; Hu, Y.; Li, Y. Molecularly imprinted polymer-decorated signal on-off ratiometric electrochemical sensor for selectiv robust dopamine detection. *Biosens. Bioelectron.* **2019**, *135*, 224–230. [CrossRef]
50. Beatto, T.G.; Gomes, W.E.; Etchegaray, A.; Gupta, R.; Mendes, R.K. Dopamine levels determined in synthetic urine usi electrochemical tyrosinase biosensor based on ZnO@Au core–shell. *RSC Adv.* **2023**, *13*, 33424–33429. [CrossRef]
51. Sun, X.; Zhang, L.; Zhang, X.; Liu, X.; Jian, J.; Kong, D.; Zeng, D.; Yuan, H.; Feng, S. Electrochemical dopamine sensor bas superionic conducting potassium ferrite. *Biosens. Bioelectron.* **2020**, *153*, 112045. [CrossRef] [PubMed]
52. Li, Y.-Y.; Kang, P.; Wang, S.-Q.; Liu, Z.-G.; Li, Y.-X.; Guo, Z. Ag nanoparticles anchored onto porous CuO nanobelts fc ultrasensitive electrochemical detection of dopamine in human serum. *Sens. Actuators B Chem.* **2021**, *327*, 128878. [CrossR
53. Liang, Y.; Guo, T.; Zhou, L.; Offenhäusser, A.; Mayer, D. Label-Free Split Aptamer Sensor for Femtomolar Detection of Dopa by Means of Flexible Organic Electrochemical Transistors. *Materials* **2020**, *13*, 2577. [CrossRef] [PubMed]
54. Tiwari, J.N.; Vij, V.; Kemp, K.C.; Kim, K.S. Engineered Carbon-Nanomaterial-Based Electrochemical Sensors for Biomole *ACS Nano* **2016**, *10*, 46–80. [CrossRef] [PubMed]
55. Anithaa, A.C.; Lavanya, N.; Asokan, K.; Sekar, C. WO_3 nanoparticles based direct electrochemical dopamine sensor i presence of ascorbic acid. *Electrochim. Acta* **2015**, *167*, 294–302. [CrossRef]
56. Arvand, M.; Ghodsi, N. Electrospun TiO_2 nanofiber/graphite oxide modified electrode for electrochemical detection of l-D in human cerebrospinal fluid. *Sens. Actuators B Chem.* **2014**, *204*, 393–401. [CrossRef]

molecules

MDPI

iew

anomaterials-Based Electrochemical Aptasensors for Rapid etection of Pathogens and By-Products

ang Lei [1] and Norjihada Izzah Ismail [1,2,*]

1 Bioinspired Device and Tissue Engineering (BIOINSPIRA) Research Group, Department of Biomedical Engineering and Health Sciences, Faculty of Electrical Engineering, Universiti Teknologi Malaysia, Johor Bahru 81310, Johor, Malaysia; zhanglei@graduate.utm.my

2 Medical Devices and Technology Centre, Institute of Human Centered Engineering, Universiti Teknologi Malaysia, Johor Bahru 81310, Johor, Malaysia

* Correspondence: norjihada@utm.my

Abstract

The rapid and accurate detection of pathogenic bacteria and viruses is essential for controlling infectious disease outbreaks and ensuring food safety. Conventional detection methods such as microbial culture, immunoassays, and polymerase chain reaction (PCR), although effective, often suffer from drawbacks including time-consuming procedures, complex operations, and limited multiplexing capabilities. In recent years, electrochemical aptasensors have emerged as a promising alternative for rapid detection of pathogenic bacteria, viruses, and by-products (toxins) due to their high sensitivity, excellent specificity, low cost, and potential for miniaturization. Aptamers can be applied as biorecognition elements of the biosensor, remarkably offering advantages such as high binding affinity, thermal stability, and ease of chemical synthesis. Meanwhile, nanomaterials which provide large surface area, superior conductivity, and modifiable surfaces are widely employed in signal amplification and sensor platform construction. This review discusses the cutting-edge innovations in electrochemical aptasensors in recent years that utilize various types of nanomaterials to accurately identify and quantify diverse types of pathogens and toxins. This review focuses on nanomaterials such as metal nanostructures, carbon nanomaterials, metal, metal oxides, and carbon nanocomposites that can synergistically enhance detection sensitivity, specificity, and operational stability. This review also highlights the promising practical application of the proposed electrochemical aptasensors in clinical diagnostics, environmental monitoring, and food safety.

Keywords: electrochemical aptasensor; aptamer; rapid detection; nanomaterials; pathogen; toxin; bacteria; virus

1. Introduction

Pathogens such as bacteria and viruses spread at extremely fast speeds, posing unprecedented challenges to global public health and cross-border epidemic prevention and control [1,2]. Specifically, bacteria are one of the main sources of epidemics and are mainly found in contaminated water, food, and other biological samples, making them contagious [3,4]. History has also shown that viruses can easily acquire the ability to initiate and spread a pandemic due to their rapid spread and high rate of transmission of new mutations, as well as the difficulty faced by humans in obtaining rapid and accurate diagnostics, specific vaccines, or therapeutic drugs in a short period of time [5,6].

While rapid, sensitive, and specific detections of pathogenic bacteria and viruses are still challenging, techniques involving cell culture, immunological reaction, polymerase chain reaction (PCR) and mass spectrometry are the current mainstream detection methods [7–12]. Since cell culture-based methods include selective enrichment, biochemical screening, and humoral confirmation steps, it takes 24–48 h to obtain preliminary results of identification, which is time-consuming and labor-intensive [7,13,14]. Immunoassays such as enzyme-linked immunosorbent assay (ELISA) have been widely used for pathogen detection. ELISA has been proven to be the gold standard for detecting *Staphylococcus aureus* enterotoxins in food production facilities [8,10]. However, three key limitations restrict its application. The assay requires labor-intensive steps (e.g., repeated washing, incubation), which prolongs the processing time by 2–4 h compared to the automated methods. In addition, obtaining reliable results typically require 50–100 μL of sample, which is challenging in situations where sample scarcity exists (e.g., neonatal blood). ELISA has sensitivity limitations in which the detection thresholds are limited to the nanomolar range (≥1 nM), therefore preventing detection of early biomarkers such as interleukin-6 in sepsis (0.1–0.5 nM) [15–17].

PCR technology can improve sensitivity (attomole detection) and specificity, shortening the diagnostic cycle to 1–24 h. Multiplex platforms of PCR enable simultaneous detection of bacteria such as *S. aureus* and *Staphylococcus epidermidis* within 90 min [12]. Real-time reverse-transcriptase PCR (qRT-PCR) is able to detect 14 respiratory viruses including SARS-CoV-2 and influenza A virus with clinical consistency up to 95% [18]. Despite these advantages, several disadvantages remain. In terms of pretreatment, a 12–24 h pre-enrichment step is usually required to amplify bacterial DNA and to exclude inactive pathogens [19]. The other issue is linked to matrix interference. Complex samples (e.g., meat homogenates) normally contain PCR inhibitors (e.g., polysaccharides, fats) which can reduce the amplification efficiency by 30–60% unless the samples are pretreated by centrifugation or filtration [9,20].

Biosensors have been introduced as another strategy for pathogen detection and are notably preferred due to several advantages of these instruments: highly sensitive and selective, rapid, easy sample preparation, are easy to use, allow on-site detection and are cost-effective [21,22]. The biosensor architecture integrates three core modules: (1) biorecognition (e.g., engineered antibodies/aptamers), (2) transducer (e.g., nanomaterial-enhanced electrodes), and (3) signal processor (i.e., microchip-embedded algorithms) to achieve automated quantitative analysis of pathogens [23]. Biorecognition is the specific interaction of antibodies or aptamers with target analytes, which improves the sensitivity and selectivity of biosensors, achieves high-precision detection, and minimizes other interferences in the sample [21].

In the past 10 years, aptamer has gained enormous attention for use in biosensor development. A study revealed that the carbon nanotube (CNT)-based field effect transistor (FET) biosensor achieved highly sensitive detection of foodborne pathogens *Salmonella* enterica and *S. aureus* through aptamer functionalization. The aptamer biosensor (aptasensor) maintained a stable performance in six complex food matrices, with a detection limit as low as 3.1 CFU, a response time as short as 200 s, and no cross-reaction with closely related bacteria such as *Listeria* [24]. Another study developed an electrochemical aptasensor for the detection of *Acinetobacter baumannii* with a detection limit of 1 CFU/mL. The functional groups, namely, hydroxyl and carboxyl groups in the synthetic carbon quantum dots (CQD), reduced graphene oxide (rGO), multi-walled carbon nanotubes (MWCNTs), and chitosan (CS) nanocomposite help immobilize a large number of aptamers onto the electrode which effectively enhanced the stability and activity of the aptasensor. This nanocomposite material exhibited high conductivity and fast electron-transfer kinetics

https://doi.org/10.3390/molecules310

which promote enhanced sensitivity and specificity. In addition, the large surface area offered by the hemin-graphite oxide (H-GO) attached to the secondary aptamers increased the aptamer surface density, leading to signal amplification and better sensitivity. This aptasensor was used to monitor the presence of live *A. baumannii* cells in human serum and skimmed milk powder, working as a platform for early detection and diagnosis of this pathogen-associated infection [25].

2. Electrochemical Biosensors and Aptamers for Bacterial and Viral Detection

Electrochemical biosensors convert biochemical signals into electrical signals through electrodes and provide output in qualitative and quantitative information of target analytes after processing [26]. Compared with traditional methods, the advantage of electrochemical biosensors is that they can immobilize aptamers to form highly sensitive electrochemical aptasensors. Aptamers can be fixed to the surface of gold or carbon-based electrodes through chemical bonds. When combined with nanostructured amplification strategies, these platforms achieve single-cell-level detection in bacterial detection and exhibit picomolar to femtomolar limits of detection in viral detection while maintaining specificity in complex sample media such as food, water, and serum [27].

Aptamers are synthetic single-stranded oligonucleotides (DNA or RNA) developed through the systematic evolution of ligands by exponential enrichment (SELEX) process and are commonly 25–80 nucleotides in length [20,27,28]. Aptamers can fold into a defined three-dimensional structure following intramolecular base-pairing which results in their high affinity and specificity to bind with targets such as small molecules, toxins, bacteria, and viruses [27–29]. The aptamers can fold into hairpin, loops, stem, bulges, and G-quadruplexes for molecular recognition and non-covalent interactions such as hydrogen bonding, polar contacts, van der Waals forces, hydrophobic interactions, stacking of aromatic rings and electrostatic interactions mediate the aptamer–target binding [28]. Since their first report in 1990, aptamers have been widely studied due to their superior properties compared to antibodies including non-immunogenicity, high specificity, easy modification, stability, shorter production time, lower cost, and scalable synthesis [29,30]. Interestingly, aptamers have been shown to be capable of differentiating thousands of proteins and nucleotides within a short time as well as identifying small variations between proteins with similar structures [29].

SELEX is an aptamer screening strategy which employs a series of selection and amplification steps. It starts with screening a random library of single-stranded DNA or RNA against the specific target under specific conditions such as a specific temperature and salt concentration. The nucleic acid molecules bound with the target of interest are isolated from the unbound targets, and PCR amplification is used to selectively enrich these target-specific aptamers, followed by high-throughput sequencing [28,30]. Isolation of aptamers that bind specifically to the target may require 1 to 20 rounds of SELEX depending on the nature of target [28]. While the SELEX process has been time-consuming in the past decade, improvements in SELEX technology through in silico pre-screening have shortened the aptamer development process and reduced the production costs [30–33]. With high-efficiency SELEX technology, pathogen-specific aptamers can be rapidly obtained during new epidemic outbreaks, such as the recent acquisition of specific aptamers for severe acute respiratory syndrome coronavirus 2 (SARS-CoV-2) [34–36].

Aptamers can identify pathogens by targeting surface antigens such as lipopolysaccharide or unknown structures. Table 1 lists several aptamers and their targets. At present, numerous bacterial aptamers are available, primarily target pathogens such as *S. aureus*, *Listeria monocytogenes*, *Salmonella* spp., *Escherichia coli*, *Streptococcus pyogenes*, *Mycobac-*

https://doi.org/10.3390/molecules31040664

terium tuberculosis, and *Pseudomonas aeruginosa*. The aptamers of various lengths have been used [33,37–45]. It was also highlighted in the past study that the MPT64 aptamer offers more sustainable, stable, and low-cost alternatives to antibodies in the development of point-of-care biosensors, allowing shorter detection time of 30 min. In addition, MPT64-specific aptamers enable sensitive electrochemical impedance detection of secretory antigens of *M. tuberculosis* at a limit of detection of 81 pM without the need for fluorescent or enzyme labeling [38].

Table 1. Aptamers for detection of target bacteria and viruses.

Aptamer	Sequence of Aptamers	Target Bacteria/ Virus	Associated Disease	Refere
AP_7462DNA	5′-GCGGCGCGGTATGGAATTAGTGACTTCC GCGCGCCCCATTTTTTATAGGGGCCGC-3′	SARS-CoV-2 virus (spike protein)	COVID-19	[33
SA43	5′-SHTCGGCACGTCAGTAGCGCTCGC TGGTCATCCCACAGCTACGTC-3′	*S. aureus*	Septicemia, sepsis, skin and soft tissue infections, endocarditis	[37
Thiolated MTP64 aptamer	HS-(CH6)6-OP(O)2O-(CH2CH2O)6-5′-TTTTT-aptamer-3′	*M. tuberculosis* (MPT64 protein)	Tuberculosis	[38
6011	5′-GCCTGTTGTGAGCCTCCTGTCGAAC AACCACTCATATCTACTACATGACTTG CTCCATTCTGTTCTTTCTCTACGCATTG AGCGTTTATTCTTGTCTCCC-3′	Zaire Ebola virus (EBOV) soluble glycoprotein (sGP) and surface glycoprotein (GP1,2)	Hemorrhagic fever	[39
LM6-116	5′-AGTATACGTATTACCTGCAGCTACTC GTTATTTCGTAGCACTTTTCCCCACCAC CTTGGTGCGATATCTCGGAGATCTTGC-3′	*L. monocytogenes*	Sepsis, meningitis	[40
apt 1; apt 2	5′-SH-AGTAATGCCCGGTAGTTATTCA AAGATGAGTAGGAAAAGA-3′ (apt 1); 5′-ROX-AGTAATGCCCGGTAGTTATTCAAA GATGAGTAGGAAAAGA-3′ (apt 2)	*Salmonella typhimurium* (whole cell)	Gastroenteritis	[41
E-CA 20; E-CA 20P	5′-CACACACGGAACCCCGACAACATA CATACGGTGAGGGTGG-3′; 5′-TTCACGGTAGCACGCATAGGCACAC ACGGAACCCCGACAACATACATACGG TGAGGGTGGCATCTGACCTCTGTGCTGCT-3′	*Streptococcus pyogenes* (M-type, M11 protein)	Pharyngitis, necrotizing fasciitis, sepsis, rheumatic fever, glomerulonephritis	[42
P12-21, P12-55; P12-11, P12-31; P12-52a, P12-52b (clone)	CCGGAGGTGGGTGAGGTCTGCGGCAGG CTGTGTGGGTGGACCGGAGGGGGGTGAG-GTCTGCGGCAGGCTGTGTGGGTGGA; CCCTCCGGGGGGGGGGTCATCGGGATA CCTGGTAAGGATACCCTCCGGGGGGG ----TCATCGGGATACCTGGTAAGGATA; CCGCCCAGCGGGGGTAGGGCCGGACG TAGGAGGAGCTGCG	*E. coli*, (whole cell)	Urinary tract infections, intestinal and diarrheal diseases, sepsis/meningitis	[43
JN27	5′-ATGAGAGCGTCGGTGTGGTAACTAGTC TGATTTCTATTTCCTTTAATTAGTCTGCAC ACATTGCATTGTAGGAGGGTGCGGAAGTA-3′	*P. aeruginosa*	Cystic fibrosis, infections in burn wounds	[44

https://doi.org/10.3390/molecules31

Table 1. *Cont.*

tamer	Sequence of Aptamers	Target Bacteria/ Virus	Associated Disease	Reference
V ssDNA tamer	5′-NH2-GGGGGGCCAAGGCCCAGCCC TCACACA-3′	Human immunodeficiency virus (HIV)-1	Acquired immune deficiency syndrome (AIDS)	[45]
CVcp ssDNA tamer	-	Hepatitis C virus (HCV) (core protein)	Liver cirrhosis, hepatocarcinoma	[46]
Apt-Fc	5′-AGTATACCGTATTACCTGCAGCCATGTTT TGTAGGTGTAATAGGTCATGTTAGGGTTT CTGCGATATCTCGGAGATCTTGC-3′	Norovirus	Gastroenteritis	[47]
	5′-GGCTGTTGTTGTTACCTATTGCGTGGCG ATCGGACTTTCGATTCCGATTAACGCCGG AGG-3	Zika virus (NS1 protein)	Guillain–Barré syndrome (GBS), meningoen-cephalitis	[48]

Viral pathogens including human immunodeficiency virus (HIV-1), hepatitis B virus (HBV), and SARS-CoV-2 are increasingly being targeted by aptamer-based detection platforms. Cossettini and colleagues [33] has computationally designed 18 high-affinity aptamers (Kd < 1.2 nM) for the SARS-CoV-2 spike receptor binding domain using a deep learning-accelerated SELEX workflow (AptaNet v2.0) and achieved high specificity in distinguishing Omicron subtypes (BA.5 vs. XBB.1.5) using electrochemical impedance spectroscopy. For detection of glycoprotein (GP) and soluble glycoprotein of Ebola virus (EBOV sGP) (Table 1), a past study revealed that all three studied aptamers, namely, 6011, 6012, and 39SGP1A, exhibited similar affinities to sGP and GP1,2 proteins. One of the aptamers (6011) was selected as an electrochemical sensor element and the study confirmed that Ebola GP1,2 proteins on pseudotyped virions can be detected while high sensitivity of EBOV sGP detection was observed with limit of detection of 150 pM [39].

Pathogenic bacteria such as *Salmonella enterica* serovar *Typhi* (typhoid fever), *E. coli* O157:H7 (hemolytic uremic syndrome), *S. aureus* (toxic shock syndrome), and *Shigella dysenteriae* (bacillary dysentery) cause more than 1.5 million deaths worldwide each year, mainly in resource-poor settings [49–53]. These pathogens employ different virulence mechanisms. *Salmonella enterica* serovar *Typhi* invades intestinal epithelial cells via a type III secretion system, causing systemic infection with a mortality rate of 10–20% if untreated [50]. *E. coli* O157:H7 produces Shiga toxin which damages renal endothelial cells and may lead to acute kidney injury (AKI) progression to renal failure in pediatric cases [51]. *S. aureus* secretes various virulence factors including superantigens (e.g., toxic shock syndrome toxin-1 [TSST-1]) and causes bacteremia which results in shock or sepsis with mortality rates between 10 and 80% [52]. *Shigella* species, which is responsible for up to 165 million infections each year, colonizes the colonic mucosa using actin-based peristalsis and can cause severe bloody diarrhea [53].

Deadly viruses such as influenza viruses, Ebola virus, human immunodeficiency virus (HIV), Hantavirus and Dengue virus can affect people in a specific region or from all over the world and are therefore accounted as significant threats to human health [26,54–58]. SARS-CoV-2, the pathogen of COVID-19, caused nearly 14.83 million excess deaths globally as estimated by WHO [59]. COVID-19 has been estimated to have caused global economic losses between US $77 billion and US $2.7 trillion in 2019, and since then has increased the burden on medical infrastructure worldwide due to the elevated medical costs, healthcare use, and medications [60,61]. It is worth noting that bacterial coinfection and secondary bacterial infection are also considered as the key risk factors for the observed severity and

https://doi.org/10.3390/molecules31040664

mortality of COVID-19 patients [62–64]. The viral infection weakens the host immune system, paving a way for the development of viral–bacterial coinfections [65,66].

Most importantly, the re-emergence of viral diseases poses a continuous threat to humans and healthcare settings. Recent reports revealed that over 3000 cases of monkeypox, a rare zoonotic viral disease transmitted from animals to humans, have been recorded in more than 50 countries since early May 2022 [67]. Marburg virus (MARV) is a highly pathogenic virus that causes a filoviral hemorrhagic fever characterized by hemorrhagic manifestations and multi-organ failure, with a mortality rate of up to 88% among confirmed cases in the 2022 Marburg outbreak in Equatorial Guinea [6,68,69]. Multiple outbreaks of MVD have occurred in Africa over the past fifty years, and two outbreaks were also reported in Europe [70]. Since its discovery near the Ebola River in 1976, more than 30 discrete outbreaks of Ebola virus disease (EVD) have been documented, with clusters occurring in Central and West Africa. In 2021, Guinea experienced a new EVD outbreak [71]. The 2014–2016 West African epidemic with 28,646 cases and 11,323 deaths, and 2018–2020 Kivu outbreak with 3470 cases and 2287 deaths generated over 17,000 survivors with persistent post-Ebola sequelae (ocular/neurological complications) [57,72].

Oncogenic viruses cause 15–20% of cancer incidence worldwide; for example, human papillomavirus (HPV) causes hundreds of thousands of cases of cervical cancer each year [73,74], while HBV and hepatitis C virus (HCV) work together to cause 80% of hepatocellular carcinoma in endemic areas, and this proportion is even higher in high-endemic areas such as sub-Saharan Africa and East Asia [75]. Cancer is a significant threat to human health, causing nearly 10 million deaths globally every year, and viral infections are responsible for a considerable proportion of these deaths [76]. Therefore, the use of electrochemical aptamer sensors to continuously monitor the presence and quantity of tumor viruses has important clinical significance for early cancer diagnosis and treatment [77].

3. Nanomaterials in Electrochemical Aptasensors for Detection of Pathogens and By-Products

Electrochemical aptasensors have gained prominence for pathogen detection, considering the critical needs of rapid detection in clinical and food safety contexts [3,78]. These sensors commonly utilize nanomaterials as dual-functional platforms: immobilizing aptamers and amplifying electrochemical signals. Nanomaterials, owing to their small particle size in the nanoscale range (1–100 nm), have been utilized in development of numerous biosensing tools for a wide range of applications such as biotechnology, medical diagnostic, environment, agriculture, and food safety [79]. Figure 1 depicts the utilization of nanomaterials in the electrochemical aptasensor where the nanomaterials are deposited onto a working electrode such as a glassy carbon electrode (GCE).

Five nanomaterial classes dominate current research: metal/metal oxide nanoparticles (e.g., gold (Au), silver (Ag), magnetite (Fe_3O_4)), carbon-based materials (e.g., graphene, carbon nanotubes), quantum dots (QD), polymer nanoparticles, and hybrid composites [37,78,80]. The performance superiority of the nanomaterials-based electrochemical aptasensors arises from the nanomaterials contributions such as enhanced surface area, increased sensitivity and specificity, multiplexing, multi-analyte detection, ease of functionalization, rapid detection, real-time monitoring, reduce false positives and negatives, and consistent performance [81,82]. Interestingly, nanomaterials can be integrated into the working electrode of electrochemical sensors via various techniques including electrodeposition, surface modification, layer-by-layer assembly, sol–gel technique, and printing technologies [82].

https://doi.org/10.3390/molecules310

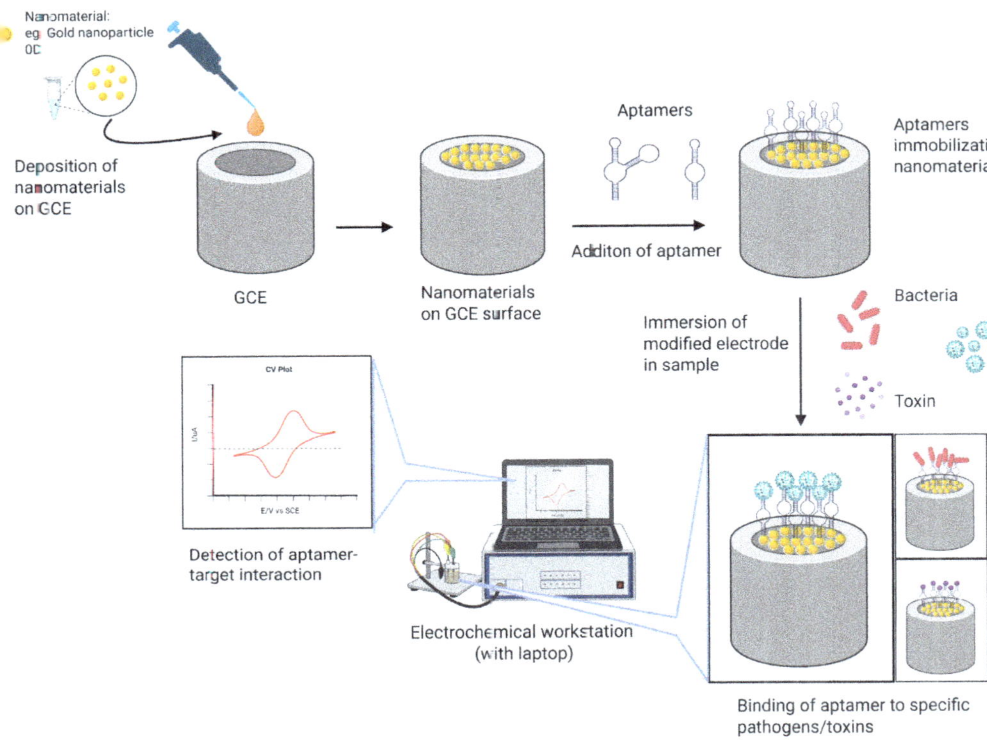

Figure 1. Modification of electrochemical sensor with nanomaterials and aptamers for pathogens and toxins detection.

As the bioreceptor of aptasensors with stable secondary structure, aptamers are well-known for their high binding affinity and selectivity for specific target molecules, such as amino acids, proteins, enzymes, and metal ions [79,81]. These properties are attributed to the ability of the aptamers to fold via electrostatic interactions, van der Waals forces and hydrogen bonds, forming thermodynamically stable three-dimensional structures. The accuracy and sensitivity of the assays can be further improved while the limit of detection can be reduced by utilizing the combination of aptamers with different nanomaterials [79]. To ensure this, aptamer conjugation with nanomaterials is a crucial step for effective functionality of aptasensors especially to preserve the short single-stranded DNA (ssDNA) or RNA biophysical characteristics and binding abilities. Several methods are available for the conjugation, namely, physical adsorption, covalent and non-covalent attachments [81]. Figure 2a presents the examples of aptamer immobilization strategies on nanomaterials for electrochemical aptasensors. After successful aptamer immobilization on the nanomaterials, subsequent binding between the aptamer and target will produce a measurable electrical signal, and alteration in electrical signals before and after binding can also be observed (Figure 2b).

https://doi.org/10.3390/molecules31040664

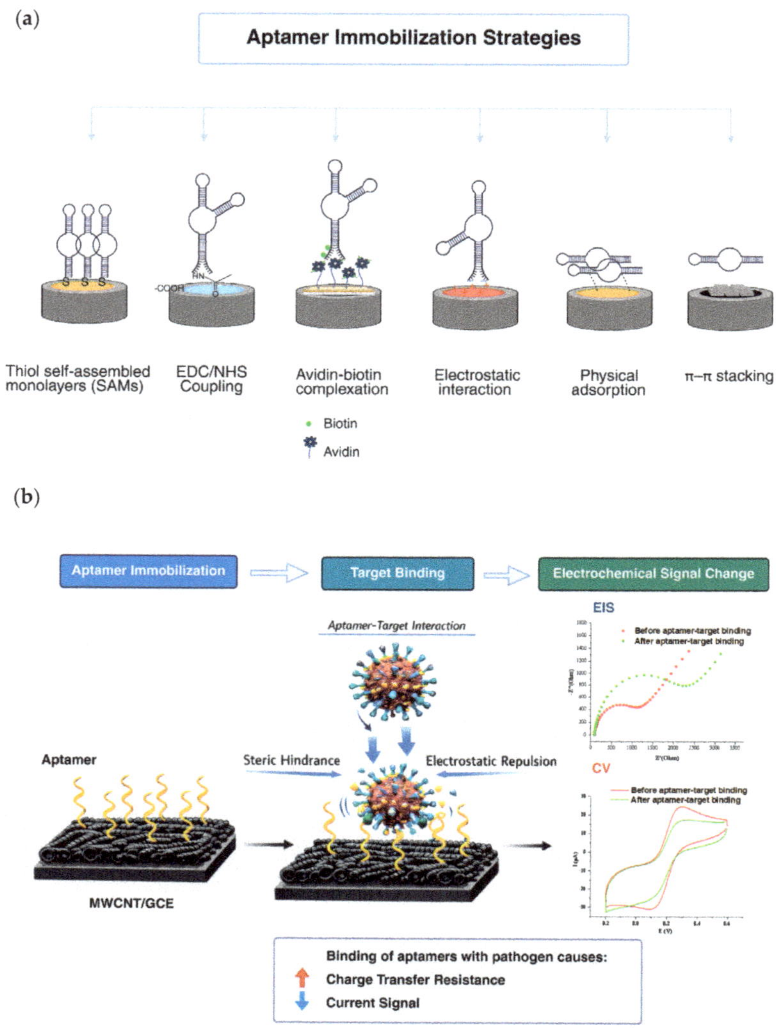

Figure 2. Schematic illustration of (**a**) strategies of aptamers immobilization on nanomaterials for electrochemical aptasensors and (**b**) binding of aptamers with the target to generate electrical signals.

3.1. Metal Nanoparticles

Metallic nanostructures are becoming the key components in the optimization of aptasensors due to their high surface-area-to-volume ratio, fast electron transfer kinetics, increased loading capacity, mass transport of reactants, and inherent cytocompatibility [79,83]. Gold (Au) and silver (Ag) nanoparticles are widely used in customized sensor design due to their excellent conductivity, and controllable size and morphology (e.g., nanospheres, nanorods, nanocages, nanowires, etc.) [79,84]. Gold nanoparticles (AuNPs) have gained worldwide attention for diverse diagnostic and therapeutic applications due to their inertness, biocompatibility, low toxicity, chemically stable, localized surface plasmon resonance (LSPR) properties and easy modification [79,85]. The conventional synthesis method of AuNPs is via reduction of gold (III) derivatives using citrate or other reducing agents such as gallic acid [86].

https://doi.org/10.3390/molecules31

Similarly, silver nanoparticles (AgNPs), which can be synthesized conventionally through chemical and physical methods, are cost-effective and have demonstrated minimal cytotoxicity and immunological response. As one of the preferred nanomaterials for electrochemical sensors, AgNPs enable rapid and sensitive detection, which supports its utilization in point-of-care devices [87]. In the past years, plant extracts and microbial (i.e., bacteria, fungi, yeasts, actinomycetes, and viruses) green synthesis methods of AuNPs and AgNPs have been developed, which was proven effective for the development of sustainable production systems [85–88]. These environmentally friendly approaches offer several advantages including the absence of toxic by-products from the chemical-reducing agents which could be introduced to the environment and pose health hazards, reduced energy demands and costs for nanoparticle production, and enhanced scalability by using non-toxic materials [86,87].

Many studies have reported successful detection of pathogenic bacteria and viruses using AuNPs-modified electrochemical aptasensors [89–96]. Zarei et al. [90] have successfully applied a strategy of thiol-bonded aptamer fixation for the detection of *S. dysenteriae*. The aptasensor utilized a glassy carbon electrode (GCE) modified with AuNPs through electrodeposition technique and the AuNP/GCE was combined with a thiolated aptamer via self-assembly immobilization to detect *S. dysenteriae* in milk and water samples. This study reported that the aptasensor has a detection limit of 100 CFU/mL (Table 2) and a recovery rate of 93.26–132.95%. The aptasensor also showed higher selectivity, in which the presence of other bacteria and dead *S. dysenteriae* did not affect its performance. Another past study revealed that metallic nanoparticles such as AuNPs can be coupled with rolling circle amplification (RCA) technology to further increase sensor sensitivity for detection of live *S. typhimurium*. The RCA products which are the massive, long DNA molecules with multiple tandem-repeat sequences can later hybridized with the detection probe on the aptasensor surface [91].

Table 2. Metal nanoparticles-modified electrochemical aptasensors for detection of pathogens and by-products in various samples [1].

rget	Material	Method	LOD	Range	Detection Time	Detection Sample	Reference
eruginosa	AuNPs	Amperometry	60 CFU/mL	60.0–6.0 × 10^7 CFU/mL	10 min	Water	[89]
gella enteriae	AuNPs	EIS	10^0 CFU/mL	10^1–10^6 CFU/mL	30 min	Water, skim milk	[90]
monella himurium	AuNPs	DPV	16 CFU/mL	20–2 × 10^8 CFU/mL	>3.5 h	Mineral water	[91]
N1 virus nagglutinin A) protein	pAuNP	CV	1 pM	1 pM–100 nM	2 h	Chicken serum	[92]
RS-CoV-2 us spike protein	AuNPs	DPV	91.1 pM	10 pM–6 nM	35 min	Artificial saliva, human serum	[93]
a virus	AuNPs	DPV	0.2 fM/ 33 fM	10–600 fM/ 500 fM–10 pM	N/A	Serum	[94]

https://doi.org/10.3390/molecules31040664

Table 2. *Cont.*

Target	Material	Method	LOD	Range	Detection Time	Detection Sample	Referen
Microcystin-LR (MC-LR), Cylindrospermopsin (CYL), anatoxin-α, saxitoxin and okadaic acid (OA)	AuNPs	SWV	0.0033 nM (MC-LR), 0.0045 nM (CYL), 0.0034 nM (anatoxin-α), 0.0053 nM (saxitoxin) and 0.0048 nM (OA)	0.073–150 nM (MC-LR) 0.018–200 nM (CYL, anatoxin-α, saxitoxin, OA)	20 min	Tap water	[95]
P. aeruginosa	AgNP	EIS	33 CFU/mL	10^2–10^7 CFU/mL	50 min	Human serum	[96]
S. aureus	AgNP	DPV	1 CFU/mL	10^1–10^7 CFU/mL	N/A	Water	[97]
Saxitoxin (STX)	AgNP	DPV	1 nM	0.04–0.15 μM	N/A	Clams, mantis shrimp	[98]
Salmonella typhimurium	NiNWs	EIS	80 CFU/mL	10^2–10^6 CFU/mL	2 h	Chicken meat	[99]

[1] AgNPs: silver nanoparticles; AuNp; gold nanoparticles; CV = cyclic voltammetry; DPV = differential pulse voltammetry; EIS = electrochemical impedance spectroscopy; N/A: not available; NiNWs: nickel nanowires; pAuNP: porous gold nanoparticles; SWV: square wave voltammetry.

Different types and/or functionalization have also been performed on AuNPs for its utilization in electrochemical aptasensor. Porous AuNPs (pAuNPs) which were synthesized by reducing the agent-assisted excessive galvanic replacement method with diameter distribution of 60–110 nm has been utilized in the fabrication of multifunctional DNA structure on pAuNPs/Au electrode for detection of hemagglutinin (HA) protein, an envelope protein of avian influenza virus (H5N1) [92]. From the cyclic voltammogram (CV) results, the authors pointed out that binding of HA protein to the bioprobe AIapt/Zyme/SH-DNA 3WJ hampered electron transfer, resulting in increasing redox peaks of AIapt/Zyme/SH-DNA 3WJ with decreasing HA protein concentration. That study also suggested that the electrode modified with pAuNPs achieve higher coverage and thus enhanced signal due to the higher surface roughness and active area the pAuNPs provide [92].

A past study revealed that their aptasensor composed of indium tin oxide (ITO) glass electrode modified with AuNPs–Cysteine–aptamer exhibited increased current intensity with an increase in SARS-CoV-2 virus spike (S) protein concentration between 10 pM and 6 nM as detected by differential pulse voltammetry technique (DPV) [93]. Roushani et al. [96] modified GCE with electrodeposited AgNPs and immobilized aptamers via covalent bonding between the aptamer amino group and AgNPs. The study successfully detected *P. aeruginosa* at the concentrations of 10^2–10^7 CFU/mL with a limit of detection of 33 CFU/mL (Table 2). The study highlighted that the presence of AgNPs on the electrode surface improved the signal due to its large surface area and fast electron transfer. The study also reported that 50 min would be suitable for target binding time while 4 h was required to allow optimal aptamer immobilization [96].

Several strategies have been implemented to further improve the performance of electrochemical aptasensors. For instance, nonspecific adsorption in complex samples and electrode biofouling can reduce sensitivity and specific bacterial detection, and these problems can be alleviated using polymers such as polyethylene glycol (PEG) as an antifouling agent [100]. In addition, dual aptamer synergistic recognition based on aptamers that target specific proteins can further optimize the sensitivity of the electrochemical aptasensors.

https://doi.org/10.3390/molecules310

For example, utilization of different aptamers that target the same MPT64 protein, which is the protein secreted in the early and middle period of *M. tuberculosis* growth, was able to further improve the sensitivity of the aptasensor supported by a low detection limit of 10 fg/mL [101]. Multiple aptamers have also been used in the development of multiplexed electrochemical aptasensors, mainly to allow simultaneous determination of substances such as five harmful marine toxins in tap water [95]

3.2. Metal/Metal Oxide Nanocomposites

Metal/metal oxide nanocomposites have been utilized to modify the working electrode in electrochemical aptasensor research studies. Nanocomposites are hybrid materials that can be composed of metallic, non-metallic, and polymeric materials. Incorporation of carbon nanotubes to metal nanoparticles, for example, CNT and AuNP, form AuNP–CNT nanocomposites which display enhanced performance and properties. The AuNP–CNT nanocomposites could provide exceptional electrical conductivity, high sensitivity and selectivity, and easy surface modification, which favors their use in sensor application [102]. Many research studies have utilized metal nanocomposites for development of electrochemical aptasensors [11,103–111]. The AuNPs/SPANI nanocomposite aptasensor developed by Gao et al. [103] used the anti-biological contamination properties of sulfonated polyaniline (SPANI) to act as an antifouling agent and combined it with the signal amplification effect of AuNPs to directly detect *S. aureus* with detection limit of 2 CFU/mL (Table 3).

Table 3. Metal nanocomposite-modified electrochemical aptasensors for detection of pathogens and by-products in various samples [1].

get	Material	Method	LOD	Range	Detection Time	Detection Sample	Reference
reus	Ag-Cs-Gr QDs/ $NTiO_2$	DPV	3.3 CFU/mL	10–5 × 10^8 CFU/mL	90 min	Human serum	[11]
reus	AuNPs/SPANI	EIS	2 CFU/mL	10–10^5 CFU/mL	N/A	Water, milk	[103]
reus	AuNPs/CNPs/ CNFs	EIS	1 CFU/mL	1.2 × 10–1.2 × 10^8 CFU/mL	N/A	Human serum	[104]
uberculosis igen 64)	GNP-C_{60}-PAn	DPV	20 fg/mL	0.02–1000 pg/mL	N/A	Human serums	[105]
li O157:H7	AuNPs/rGO–PVA	EIS	9.34 CFU/mL	9.2–9.2 × 10^8 CFU/mL	N/A	Tap water, milk, meat	[106]
cobacter ri (Hsp 60)	Hap-Ag-ZnO	SWV	0.429 nM	0.05–300 nM	20 min	Human serum	[107]
ria ocytogenes	Pt/HCNs	DPV	2 CFU/mL	10–10^9 CFU/mL	N/A	Milk, lettuce homogenate	[108]
ovirus V)	BP-AuNCs	DPV	0.28 ng/mL	1 ng/mL–10 μg/mL	30 min	Oyster	[109]
oxin cotoxin)	MoS_2-PANI-Chi-Au and rGO-TEPA-Au@Pt NRs	Chronoamperometry	1.79 fg/mL	10 fg/mL–100 ng/mL	N/A	Beer	[110]

https://doi.org/10.3390/molecules31040664

Table 3. *Cont.*

Target	Material	Method	LOD	Range	Detection Time	Detection Sample	Refere
Staphylococcal Enterotoxin A (SEA)	AuNUs/rGO	DPV	7.6 fM	25.0–950.0 fM	100 min	Milk, meat extract, human serum	[111

[1] Ag-Cs-Gr QDs/NTiO$_2$: silver nanoparticles–chitosan–graphene quantum dots/nitrogen-doped TiO$_2$ nanoparticles; AuNUs/rGO: gold nano urchins/reduced graphene oxide; AuNPs/CNPs/CNFs: gold nanoparticles/carbon nanoparticles/cellulose nanofibers nanocomposite; AuNPs/SPANI: gold nanoparticles/sulfonated polyaniline; BP-AuNCs: phosphorene–gold nanocomposites; GNP-C60-PAn: gold nanoparticles decorated with coil-like fullerene-doped polyaniline; Hap-Ag-ZnO: hydroxyapatite–silver–zinc oxide; MoS2-PANI-Chi-Au: molybdenum disulfide–polyaniline–chitosan–gold nanoparticles; N/A: not available; Pt/HCNs: Platinum nanoparticles/hollow carbon nanospheres; rGO-TEPA-Au@Pt NRs: Reduced graphene oxide–tetraethylene pentamine–gold@platinum nanorods.

Ranjbar and Shahrokhian [104] fabricated an electrochemical aptasensor using gold nanoparticles/carbon nanoparticles/cellulose nanofibers nanocomposite (AuNPs/CNPs/CNFs) for sensitive and selective detection of *S. aureus*. The nanocomposites exhibited excellent conductivity, good biocompatibility, and high surface area, which contributed to high sensitivity with a limit of detection of 1 CFU/mL for the aptasensor and precise detection of *S. aureus* in the spiked human serum (Table 3). Hydroxyapatite (Hap) has been used in combination with Ag, zinc oxide (ZnO) and conductive polymer polythiophene (PP) for detection of *H. pylori* heat shock protein (HSP 60) [107]. They pointed out that the multiple functional group sites of Hap-Ag-ZnO-PP composites facilitated the aptamer immobilization through the aptamer's NH_2 group. A prominent decrease in current and increase in resistance was evident following hybridization of aptamer and Hsp 60, indicating successful attachment of Hsp 60 with the aptamer that formed a barrier layer which blocks the transfer of electrons. Their proposed aptasensors showed high sensitivity with detection limit of 0.429 nM (Table 3).

A 3D electrochemical aptasensor containing phosphorene–gold nanocomposites (BP-AuNCs) was developed by Jiang et al. [109] for detection of norovirus (Table 3). They fabricated the aptasensors using pins and fabric cloth which are low cost, and the incorporation of multiple layers of BP-AuNCs has resulted in ultrasensitive detection of the recombinant norovirus-like particles (VLP) with a limit of detection of 0.28 ng/mL. The sensitivity of the aptasensor could be attributed to the BP-AuNCs which provide a durable 3D support structure for the target-specific aptamer immobilization while enhancing the electron transfer process at the interface. The study also pointed out that the aptasensor demonstrated high specificity for norovirus; current responses of interfering astrovirus and rotavirus were close to blank while the current responses of the virus mixture was comparable to that of only norovirus. It was observed that the recovery from the oyster spiked samples was 97–106%, supporting its potential use for detection in food samples [109].

Metal oxide nanoparticles such as cerium oxide (CeO_2), zinc oxide (ZnO), iron oxide (Fe_3O_4), manganese dioxide (MnO_2) and titanium dioxide (TiO_2) are attractive fundamental building blocks for advanced electrochemical aptasensor architectures. Notably, metal oxide NPs have a simple preparation process, easy size, shape and porosity modification, high stability, no swelling variations, easy functionalization which is made possible due to the negative surface charge, and easy incorporation in both hydrophilic and hydrophobic systems [112]. On top of that, some metal oxide NPs such as MnO_2 and Fe_3O_4 are favored for diagnostic purposes due to their non-toxicity and strong peroxidase-like activity, as well as low production cost as compared to natural enzymes production [113]. Specifically, nanomaterials with intrinsic enzyme-like activities are known as nanozymes. Nanozymes encompass a large number of nanomaterials which can be synthesized and display excellent

https://doi.org/10.3390/molecules31

catalytic activities as they mimic natural enzymes' structures and functions (e.g., hydrolase, peroxidase, catalase, oxidase, etc.) [114].

Sharma et al. [115] synthesized ZnO nanorods by thermal method while AuNPs and AgNPs were prepared chemically. The Ag-Au-ZnO/origami-based electrochemical paper-based analytical device (oPAD) was used to fix the CHIKV aptamer onto it, forming an aptamer/Ag-Au-ZnO/oPAD. This aptasensor has shown ultrasensitive detection of Chikungunya virus antigen (CHIKV-Ag) with a linear range of 1 ng/mL–10 µg/mL (Table 4), and a detection limit of 1 ng/mL using CV. From the field emission scanning electron microscopy (FESEM) images, the spherical Au-Ag NPs were seen clustering on the surface of ZnO nanorods, which favorably contribute to a high sensitivity of 1.36 µA $\mu gmL^{-1}cm^{-2}$ as observed for this aptasensor. This improved sensitivity could be attributed to the enhanced charge-transfer properties provided by the tertiary nanocomposite. The aptasensor exhibited rapid detection, high stability, high sensitivity to serum sample, good reproducibility, and provided confidentiality in the form of a closed environment that protects electrodes from dust and external elements [115].

Table 4. Metal/metal oxide nanocomposite-modified electrochemical aptasensors for detection of pathogens and by-products in various samples [1].

get	Material	Method	LOD	Range	Detection Time	Detection Sample	Reference
ikungunya us antigen HIKV-Ag)	Ag-Au-ZnO	CV	1 ng/mL	1 ng/mL–10 µg/mL	20 s	Human serum	[115]
aureus	$AuNPs@Fe_3O_4$	DPV	1 CFU/mL	10^1–10^7 CFU mL^{-1}	N/A	Milk, conduit water, apple juice	[116]
toxin	Au/(Ce-In)Ox	DPV	7.6×10^{-8} ng/mL	5.0×10^{-7} ng/mL–5.0 ng/mL	N/A	Maize	[117]
hratoxin A	ZnO-Au	DPV	0.05 pg/mL	0.1–30,000 pg/mL	N/A	Wine and beer	[118]

[1] Ag-Au-ZnO: silver nanoparticles–gold nanoparticles–zinc oxide nanorods; Au/(Ce-In)Ox: gold nanoparticles/bimetallic oxide (cerium–indium); $AuNPs@Fe_3O_4$: gold nanoparticles modified magnetic nanoparticles; CV: cyclic voltammetry; DPV: differential pulse voltammetry; N/A: not available; ZnO-Au: Zinc oxide–gold nanoparticles.

Another past study by El-Wekil and colleagues [116] developed an aptasensor by immobilizing aptamer on gold nanoparticles modified magnetic nanoparticles (apt-$AuNPs@Fe_3O_4$) (Table 4). In the presence of *S. aureus* as the target bacteria, a molecular imprinted polymer (MIP) film using o-phenylenediamine was electro-polymerized on the surface of the as-synthesized nanocomposite to fabricate a MIP-apt-$AuNPs@Fe_3O_4$/GCE. The *S. aureus* was later removed to allow binding of the *S. aureus* in complex matrices on the aptasensor at the imprinted sites. Their study revealed that the MIP-aptasensor demonstrated an ultrasensitive detection with a low limit of detection of 1 CFU/mL and a wide linear range of 10^1–10^7 CFU/mL. The specificity studies using CV were conducted in the presence of organic compounds such as urea, glucose, ascorbic acid, uric acid, methionine, glycine, alanine, lysine, arginine, and lactic acid (300 µM each), as well as other bacteria and *C. albicans* at a concentration of 10^6 CFU/mL. They found out that reduced current was only observed in the presence of *S. aureus*, indicating the good interaction of aptamer and well-fitted MIP spaces for the *S. aureus* [116].

3.3. Carbon Nanomaterials and Nanocomposites

Carbon-based nanomaterials, which include carbon nanotubes (CNTs), graphene, graphene oxide, nanographite, fullerene, etc., are widely used as the core materials of electrochemical aptasensors. Their unique properties including high electron mobility, good electrical conductivity, large specific surface area, high stability, biocompatibility, hydrophilicity, and flexible interface functionalization capabilities have attracted their use for various applications such as in the field of medicine, electronics, environmental monitoring and food safety [29,79,81]. CNTs are composed of single-walled carbon nanotubes (SWCNTs) and multi-walled carbon nanotubes (MWCNTs). The SWCNT is a single tube with a common diameter of 0.4–2 nm. The MWCNTs which consist of several concentric tubes normally are having a diameter of 2–100 nm and the sizes are directly influenced by their synthesis method [119]. CNTs are known for their exceptional mechanical, electrical, and thermal properties, rendering them a suitable platform for the transducer component in aptasensors. The tensile strengths exceeding 100 GPa allow CNTs to withstand mechanical stresses and harsh environmental conditions while their good electrical conductivity ranging from 10^3 to 10^4 S/cm facilitate sensitive and rapid signal transduction [81].

Graphene can easily be functionalized and exhibits excellent electrical, electrochemical, and physicochemical properties. It is also highly sensitive to external stimulus and permits ssDNA binding through non-covalent π–π interaction or electrostatic interaction with DNA bases which promotes biomolecule immobilization [120]. Graphene oxide (GO), a derivative of graphene, is rich in oxygen functional groups such as hydroxyl and carboxyl groups [83]. Both graphene and GO demonstrate excellent optoelectronic properties and can detect biomolecules with high sensitivity and selectivity [79,104]. GO is commonly transformed to reduced graphene oxide (rGO) for use in sensor application, where the conversion can be achieved by the hydrothermal and chemical reduction of GO [83]. The higher hydrophilicity, higher current density, superior electrocatalytic property, and easier functional groups attachments, such as those for aptamer immobilization or for binding to a transducer or support surface, are the among the advantages offered by rGO [83,119]. However, there are some limitations of rGO where it is prone to aggregating or possibly restacking to graphite following van der Waals forces and intense layering, which may limit its application in electrochemical sensing due to a reduction in total surface area [83].

Research studies have been conducted using carbon nanomaterials and aptamers for electrochemical detection of pathogens. Kaur et al. [121] constructed a hierarchical nanostructure of 3D bridged rebar graphene (BRG) using MWCNTs which was modified by the chemical facilitated unscrolling method, followed by bridging with terephthalaldehyde (TPA). The BRG showed enhanced electrical properties supported by the eleven-fold increase in the current and facile chemical functionality and was thus functionalized with polyL-lysine (PLL) and aptamer, forming an aptamer-functionalized BRG (BRG/PLL/Apta) for detection of *E. coli* O78:K80:H11. The work has successfully achieved ultrasensitive detection of *E. coli* O78:K80:H11 with a limit of detection of 10^1 CFU/mL and detection range of 10^1–10^6 CFU/mL in water and other samples (Table 5). It was also pointed out that there was minimal interference in the presence of other bacterial species such as *L. monocytogenes*, *Bacillus subtilis*, *Proteus vulgaris*, *E. coli* DH5α, etc., indicating the high specificity of the electrochemical aptasensor. The interaction of aptamer and *E. coli* O78:K80:H11 led to high specificity of the aptasensor, and this result was supported by their scanning electron microscopy (SEM) and confocal microscopy images. The charge transfer resistance (Rct) values showed an increasing trend with increase in bacterial concentration in the range of 10^1–10^6 CFU/mL, highlighting successful *E. coli* O78:K80:H11 binding with aptamers on the sensor surface. The aptamer–target binding on the electrode surface prevented electron transfer from the redox probe which was attributed to the steric hindrance and insulating layer formed by the large bacterial cells.

https://doi.org/10.3390/molecules310

Table 5. Carbon nanomaterials-modified electrochemical aptasensors for detection of bacteria in various samples [1].

et	Material	Method	LOD	Range	Detection Time	Detection Sample	Reference
i O73:K80:H11	BRG	EIS	10^1 CFU/mL	10–10^6 CFU/mL	≤8 min	Water, juice, and milk	[121]
onella enteritidis/ *onella typhimurium*	MWCNTs	EIS	5.5×10^1 CFU/mL/ 6.7×10^1 CFU/mL	5.5×10^1–5.5×10^6 CFU/mL/ 6.7×10^1–6.7×10^5 CFU/mL	20 min	Raw chicken meat	[122]

[1] BRG: bridged rebar graphene, EIS: electrochemical impedance spectroscopy; MWCNTs: multi-walled carbon nanotubes.

Another study utilized an amino-modified aptasensor fabricated using MWCNTs deposited on ITO electrode (ssDNA/MWCNT/ITO) for detection of pathogenic *Salmonella* bacteria [122]. The MWCNTs were electrodeposited onto the ITO electrode at 30 V for 1 min followed by functionalization of MWCNT surface using EDC-NHS chemistry. The ssDNA/MWCNT/ITO was formed by dropping the amino-modified *Salmonella* DNA aptamer onto the activated electrode. Their study revealed that a stable and regular dispersion of MWCNTs on the ITO electrode was observed from the electrodeposition process. The high specific surface area of MWCNTs promoted aptamer immobilization by binding with the reactive carboxyl groups on the MWCNTs surface and the π-π stacking between MWCNTs layers and aptamer nucleotide bases allowed binding of the aptamer on the electrode. This study identified that 5 μmol/L was the optimum concentration of DNA aptamer needed for their aptasensor development. The signal amplification effects were contributed by the MWCNTs as these nanomaterials have excellent electrical conductivity and large specific surface area. Decreased peak currents and increased impedance were evident with increasing concentrations of *Salmonella* cells. The presence of high concentration of *Salmonella* enabled more aptamer–target binding on the aptasensor surface, which reduced the effective surface area and subsequently inhibited electron transfer. Additionally, this study also compared the sensitivity limit using the PCR technique, in which the PCR resulted in a limit of detection of 10^2 CFU/mL for both *Salmonella enteritidis* and *S. typhimurium* while their aptasensor worked exclusively sensitive with a detection limit of 10^1 CFU/mL (Table 5), suggesting its suitable application for food samples.

Carbon-based nanocomposites emerge as another pivotal material in electrochemical sensing. Pathania et al. [123] synthesized MoS_2-rGO nanocomposites for the electrochemical aptasensor development which contributed to enhanced conductivity and easy biomolecular functionalization. The attachment of anti-Vi antigen aptamers on the MoS_2-rGO nanocomposite through thiol linkage enables sensitive and specific detection of *S. Typhi*, differentiating it from other *Salmonella* serovars and enteric pathogens. The advantages of using MoS_2 was related to the enhanced electrical, optical, and thermal properties as well as the tunable band gap. Their study integrated MoS_2 with rGO to prevent restacking of MoS_2 and improve the electrical conductivity. The aptasensor achieved a limit of detection of 100 pg/mL (Table 6) and high specificity was observed through selective binding of aptamer and Vi-expressing *S. Typhi* cells in the presence of different polysaccharides of other bacterial origin as supported by the confocal images. This study has successfully validated the aptasensor using urine and sera specimens spiked with Vi antigen. It was suggested that this aptasensor could be useful to track carriers of *S. Typhi* and evaluate disease prognosis, providing an easy-to-perform, rapid and reliable diagnostic test [123].

https://doi.org/10.3390/molecules31040664

Table 6. Carbon nanocomposites-modified electrochemical aptasensors for detection of pathogens and by-products in various samples [1].

Target	Material	Method	LOD	Range	Detection Time	Detection Sample	Refer
Acinetobacter baumannii	rGO/MWCNT/CS/CQD	DPV	1 CFU/mL	10–1 × 10^7 CFU/mL	N/A	Serum, Skimmed milk	[25
S. aureus	NCNO/AuNPs	EIS	3 CFU/mL	10^1–10^8 CFU/mL	15 min	Human serum	[37
M. tuberculosis	RGO/PNE/Au	LSV	0.1 × 10^{-7} μM	0.1 × 10^{-2}–0.1 × 10^{-7} μM	5 s	N/A	[83
Salmonella Typhi (Vi antigen)	MoS_2-rGO	SWV	100 pg/mL	0.1–1000 ng/mL	N/A	Sera and urine	[12
M. tuberculosis (antigen ESAT-6)	P-MOF-rGO/Pt@Au	CV	3.3 × 10^{-5} ng/mL	1.0 × 10^{-4}–2.0 × 10^2 ng/mL	1 h	Human serum	[12
Salmonella Typhimurium	rGO-AP	DPV	10^1 CFU/mL	10^1–10^8 CFU/mL	N/A	Chicken meat	[12
Salmonella Typhimurium	rGO-CHI	DPV	10^1 CFU/mL	10^1–10^6 CFU/mL	N/A	Chicken meat	[12
Salmonella Typhimurium	rGO-TiO_2	DPV	10^1 CFU/mL	10^1–10^8 CFU/mL	1 h	Chicken meat	[12
Salmonella Typhimurium	rGO-CNT	DPV	10^1 CFU/mL	10^1–10^8 CFU/mL	10 min	Chicken meat	[12
E. coli (LPS)	rGO/AuNPs	EIS	30 fg/mL/ 1 fg/mL	N/A	35 min	Human serum	[12
Lipopolysaccharide (LPS)/endotoxin	MRGO-Au	SWV	4 fg/mL/ 0.2 fg/mL	0.1–0.9 pg/mL/ 0.01–0.09 pg/mL	>35 min	Human blood serum	[13
Hepatitis B virus surface antigen (HBsAg)	rGO-AuNPs	CV	0.0014 fg/mL	0.125–2.0 fg/mL	N/A	Human serum	[13
Hepatitis C virus core antigen (Anti)	MWCNTs-Chit	DPV	1.67 fg/mL	5.0 fg/mL–1.0 pg/mL	N/A	Human serum	[13
Hepatitis C virus core antigen	3D N-C@$NiCo_2O_4$ NWs	EIS	0.16 fg/mL	0.5 fg/mL –0.12 pg/mL	N/A	Human blood serum	[13
human papillomavirus (HPV-16 L1 protein)	prGO-MoS_2	DPV	0.1 ng/mL	0.2–2 ng/mL	N/A	Human serum and saliva	[13

[1] CV: cyclic voltammetry; CS/CQD: chitosan/carbon quantum dot; DPV: differential pulse voltammetry; EIS: electrochemical impedance spectroscopy; LPS: lipopolysaccharide; LSV: linear sweep voltammetry; MoS_2-rGO: Molybdenum disulfide-reduced graphene oxide; MRGO-Au: magnetite-reduced graphene oxide with gold nanoparticles; MWCNTs-Chit: multi-walled carbon nanotubes–chitosan nanocomposite; N/A: not available; NCNO/AuNPs: nitrogen-doped carbon nano-onions/gold nanoparticles; rGO-AP: rGO-azophloxine; rGO-AuNPs: reduced graphene oxide–gold nanoparticles; rGO-CHI: reduced graphene oxide–chitosan; rGO-CNT: reduced graphene oxide–carbon nanotube; RGO/PNE/Au: reduced graphene oxide/polynorepinephrine/gold nanoparticles; rGO-TiO_2: reduced graphene oxide–titanium dioxide; prGO-MoS_2: porous reduced graphene oxide–molybdenum sulfide; Pt@Au: platinum aurum core shell nanoparticles; P-MOF-rGO: poly (diallyldimethylammonium chloride)–metal–organic framework doped reduced graphene oxide; SWV: square wave voltammetry; 3D N-C@$NiCo_2O_4$ NWs: N-doped carbon@ $NiCo_2O_4$ nanowires.

Appaturi et al. [128] hybridized rGO with carbon nanotubes (CNTs) for rapid detection of *S. typhimurium* (Table 6). The CNTs acted as one-dimensional spacers that inhibit the stacking of graphene sheets and provide large surface area, effectively increasing the loading capacity of the aptamers. The amino-modified aptamers bound covalently with the carboxyl groups of the rGO-CNT, forming amide bonds. The developed ssDNA/rGO-CNT/GCE aptasensor exhibited high sensitivity with a detection limit of 10^1 CFU/mL. The high specificity of the aptasensor could be attributed to the intermolecular folding of the aptamers that recognize only the target bacteria and prevent binding with non-*Salmonella* bacteria. The rGO-CNT contributed to enhanced electrical properties and detection of whole bacterial cell without any pretreatment or DNA extraction steps, which was made possible through the immobilization of amino-modified DNA aptamer to the nanocomposites. Differential pulse voltammetry (DPV) results revealed a decrease in oxidation current density at the potential of 0.23 V upon binding of *S. typhimurium* onto the ssDNA/rGO-CNT/GCE in a concentration-dependent manner in comparison to the ssDNA/rGO-CNT/GCE which exhibited a prominent increase in oxidation. This observed decrease in current density could be related to the bacterial cell having negative charge on its membrane which interferes and blocks electron transfer to the electrode surface [128].

Several studies have proposed electrochemical aptasensors for rapid and sensitive detection of viruses including hepatitis B (HBV) and C (HCV) viruses, and human papilloma virus (HPV) [131–134]. In a study by Rahmati et al. [133], the amino group of the aptamer promoted covalent attachment to the 3D N-C@$NiCo_2O_4$ NWs nanocomposite and the aptasensor has been utilized for HCV detection. The very high surface to volume ratio, high density of metal sites and porosity of the nanocomposite potentially enable greater aptamer load onto the GCE surface. This was evident through increase in Rct value upon aptamer immobilization (Rct = 3.08 kΩ). The addition of increasing HCV concentration further increases the Rct value, indicating restricted electron transfer due to successful formation of HCV/aptamer complex on the aptasensor surface. Their aptasensor showed high sensitivity with detection limit of 0.16 fg/mL (Table 6) with excellent recovery in real samples of human blood serum [133].

It can be summarized that diverse types of nanomaterials have been utilized in electrochemical aptasensors research, aiming at providing sensitive, specific, and rapid detection of pathogenic bacteria, viruses, and by-products (toxins). The common detection methods include CV, DPV and EIS, though other methods have also been mentioned. The detection times vary depending on factors such as sample complexity, signal amplification, aptamer stability, etc. An overview of this review is presented in Figure 3.

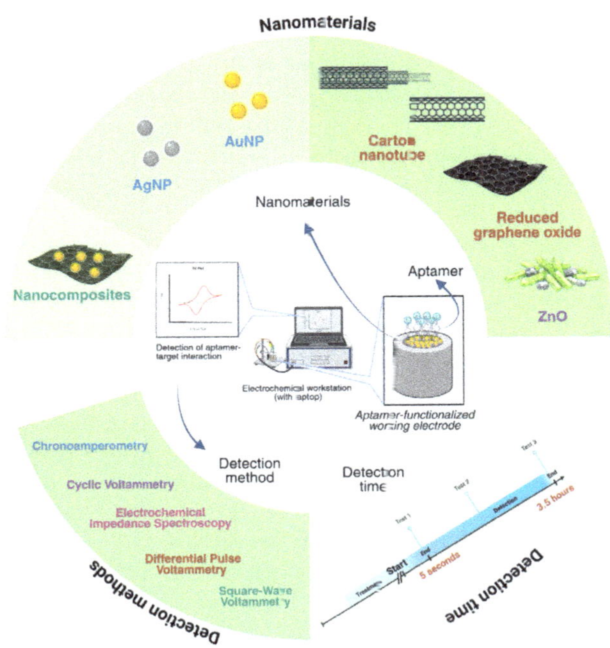

Figure 3. Overview of nanomaterials based electrochemical aptasensors that utilize different nanomaterials and methods for detection of pathogens and toxins and the time taken for the analysis.

https://doi.org/10.3390/molecules31040664

4. Challenges and Future Perspectives

The use of nanomaterials on the electrode surface to enhance signal amplification of electrochemical aptasensors helps to overcome the limitations of inhibitory effects of the aptamer on electron transfer [135]. However, it is important to note that there are limitations and challenges in using nanomaterials for electrochemical aptasensors. Aptamer immobilization which is very crucial in aptasensor development may require chemical modification or functionalization of the nanomaterials or aptamer (e.g., thiol group) for interaction between aptamer and nanomaterials [102]. In addition, the orientation and biological activity of the immobilized aptamer must be maintained as inappropriate immobilization may lead to loss of activity, low biocompatibility, and less specificity [102], which hinder the optimum interaction of target–aptamer and thus may reduce the aptasensor sensitivity. Though large DNA loading is essential for sensitive detection, it must be kept at a minimal amount to facilitate correct folding and avoid steric hindrance [136]. Other challenges would be the production cost in which the metal nanomaterials may impose more cost compared to the carbon-based nanomaterials [84]. Sensitivity of the aptasensor could also be affected by the batch-to-batch variation in nanomaterials, for example, the sizes of metal nanoparticles which result in signal fluctuations [103,137]. These challenges limit widespread clinical adoption and commercialization of nanomaterial-based biosensors for diagnostic purposes.

An ideal on-site biosensor monitoring device should meet the required sensitivity, must be highly reproducible [102], and can withstand different conditions and complexity of samples. The presence of interfering compounds known as fouling agents in complex samples such as serum also limits the functionality and effectiveness of the electrochemical aptasensors. These agents can include a wide range of molecules such as proteins, nucleic acids, amino acids, neurotransmitters, and phenols, as well as whole cells and their fragments which tend to adhere to the electrode surface through adsorption, precipitation, or polymerization. The biofouling can lead to false readings or inhibit target binding on electrode surface, which in turn affect the sensitivity, stability, reproducibility, and overall reliability of the biosensors [138]. In addition, long-term stability of aptamers remains a challenge as their activity is highly dependent on the physiological conditions. Denaturation of aptamer could potentially occur during storage, shipping, or long-term monitoring of analytes, demanding better solutions for commercial use [139]. For practical applications, electrochemical sensors must attain sensitive and specific detection as well as capability for long-term and real-time monitoring [140].

Concerns involving biofouling remain despite the existence of several strategies that are compatible with electrochemical measurements such as utilization of nanoporous gold, zwitterionic polymers, hydrophilic polymeric membranes, etc. [140,141]. Therefore, to meet the market demand of having sensitive and robust devices for analytical measurements, future research could be directed towards designing effective antifouling sensing platforms that can reduce nonspecific adsorption and increase the signal-to-noise ratio. Recent studies have also investigated the potential integration of electrochemical aptasensors with microfluidics for detection of pathogens such as norovirus [47] and *Vibrio parahaemolyticus* [142], providing advantages of high sensitivity, low cost, rapid detection, and easy miniaturization. It is noteworthy that portable diagnostic devices with high sensitivity, specificity and reproducibility are strongly desired as an alternative to the conventional methods for detection of pathogens and the by-products intended for point-of-care (POC) diagnostics, food safety, and environmental monitoring. Looking forward, integration of artificial intelligence (AI) and microfluidic-integrated lab-on-a-chip (LOC) systems in the development of electrochemical aptasensors may offer advanced analysis with improved accuracy, encourage precision medicine and personalized clinical tools, permit automation,

https://doi.org/10.3390/molecules310

easy operation and real-time monitoring as well as reduce sample volume. Considering the limitations and challenges, advancement in technology could produce a rapid detection electrochemical aptasensor diagnostic kit that may potentially help reduce diagnostic time and lower the global disease burden.

5. Conclusions

Electrochemical aptasensors utilizing nanomaterials are promising cutting-edge solutions for rapid detection of infectious disease and environmental monitoring. Taken together, through the specific binding ability of aptamer which recognizes the whole pathogen or prominent surface structures, and the large specific surface areas, stability and good electrical conductivity of nanomaterials, the high sensitivity and specificity of electrochemical aptasensors can be achieved. This review revealed that the advancement in nanomaterials technology has enabled rapid detection of infectious agents and toxins using electrochemical aptasensors. Despite the limitations and challenges that exist, electrochemical aptasensors can be considered as an alternative to the conventional methods of culture-based methods, ELISA and PCR for early infection detection, diagnosis, and monitoring. The future developments may provide better outcomes through integration of electrochemical aptasensors with AI, microfluidics, and other systems.

Author Contributions: Conceptualization, Z.L. and N.I.I.; validation, N.I.I.; data curation, Z.L. and N.I.I.; writing—original draft preparation, Z.L. and N.I.I.; writing—review and editing, N.I.I.; supervision, N.I.I. All authors have read and agreed to the published version of the manuscript.

Funding: This research was funded by Universiti Teknologi Malaysia under the grant number 04M52.

Institutional Review Board Statement: Not applicable.

Informed Consent Statement: Not applicable.

Data Availability Statement: No new data were created or analyzed in this study. Data sharing is not applicable to this article.

Acknowledgments: The authors thank Department of Biomedical Engineering and Health Sciences, Faculty of Electrical Engineering, Universiti Teknologi Malaysia for the provided facility.

Conflicts of Interest: The authors declare no conflicts of interest.

..rences

1. Zhang, Z.; Zhou, J.; Du, X. Electrochemical biosensors for detection of foodborne pathogens. *Micromachines* **2019**, *10*, 222. [CrossRef]
2. Li, H.Y.; Jia, W.N.; Li, X.Y.; Zhang, L.; Liu, C.; Wu, J. Advances in detection of infectious agents by aptamer-based technologies. *Emerg. Microbes Infect.* **2020**, *9*, 1671–1681. [CrossRef]
3. Kaur, H.; Shorie, M. Nanomaterial based aptasensors for clinical and environmental diagnostic applications. *Nanoscale Adv.* **2019**, *1*, 2123–2138. [CrossRef] [PubMed]
4. Wu, W.; Yu, C.; Wang, Q.; Zhao, F.; He, H.; Liu, C.; Yang, Q. Research advances of DNA aptasensors for foodborne pathogen detection. *Crit. Rev. Food Sci. Nutr.* **2020**, *60*, 2353–2368. [CrossRef] [PubMed]
5. Al-Rohaimi, A.H.; Al Otaibi, F. Novel SARS-CoV-2 outbreak and COVID19 disease; a systemic review on the global pandemic. *Genes Dis.* **2020**, *7*, 491–501. [CrossRef] [PubMed]
6. Sah, R.; Mohanty, A.; Reda, A.; Siddiq, A.; Mohapatra, R.K.; Dhama, K. Marburg virus re-emerged in 2022: Recently detected in Ghana, another zoonotic pathogen coming up amid rising cases of Monkeypox and ongoing COVID-19 pandemic- global health concerns and counteracting measures. *Vet. Q.* **2022**, *42*, 167–171. [CrossRef]
7. Goluch, E.D. Microbial identification using electrochemical detection of metabolites. *Trends Biotechnol.* **2017**, *35*, 1125–1128. [CrossRef]
8. Nouri, A.; Ahari, H.; Shahbazzadeh, D. Designing a direct ELISA kit for the detection of *Staphylococcus aureus* enterotoxin A in raw milk samples. *Int. J. Biol. Macromol.* **2018**, *107*, 1732–1737. [CrossRef]

9. Rajapaksha, P.; Elbourne, A.; Gangadoo, S.; Brown, R.; Cozzolino, D.; Chapman, J. A review of methods for the detecti pathogenic microorganisms. *Analyst* **2019**, *144*, 396–411. [CrossRef]
10. Shan, Y.; Xu, C.; Wang, M.; Zhu, Z.; Wu, F.-G.; Shi, Z.; Cui, Q.; Arumugam, G.M. Bilinear *Staphylococcus aureus* detection bas suspension immunoassay. *Talanta* **2019**, *192*, 154–159. [CrossRef]
11. Ghalkhani, M.; Sohouli, E.; Khaloo, S.S.; Vaziri, M.H. Architecting of an aptasensor for the *Staphylococcus aureus* analy modification of the screen-printed carbon electrode with aptamer/Ag–Cs-Gr QDs/$NTiO_2$. *Chemosphere* **2022**, *293*, 1 [CrossRef]
12. Zhou, B.; Ye, Q.; Chen, M.; Li, F.; Xiang, X.; Shang, Y.; Wang, C.; Zhang, J.; Xue, L.; Wang, J.; et al. Novel species-specific targ real-time PCR detection of four common pathogenic *Staphylococcus spp*. *Food Control* **2022**, *131*, 108478. [CrossRef]
13. Hudu, S.A.; Alshrari, A.S.; Syahida, A.; Sekawi, Z. Cell culture, technology: Enhancing the culture of diagnosing human dis *J. Clin. Diagn. Res.* **2016**, *10*, DE01–DE05. [CrossRef] [PubMed]
14. Simoska, O.; Stevenson, K.J. Electrochemical sensors for rapid diagnosis of pathogens in real time. *Analyst* **2019**, *144*, 6461 [CrossRef] [PubMed]
15. Satija, J.; Punjabi, N.; Mishra, D.; Mukherji, S. Plasmonic-ELISA: Expanding horizons. *RSC Adv.* **2016**, *6*, 85440–85456. [Cro
16. Hosseini, S.; Vázquez-Villegas, P.; Rito-Palomares, M.; Martinez-Chapa, S.O. Advantages, disadvantages and modific of conventional ELISA. In *Enzyme-Linked Immunosorbent Assay (ELISA)*; Springer Briefs in Applied Sciences and Techn Springer: Singapore, 2018. [CrossRef]
17. Wu, L.; Li, G.; Xu, X.; Zhu, L.; Huang, R.; Chen, X. Application of nano-ELISA in food analysis: Recent advances and chall *TrAC Trends Anal. Chem.* **2019**, *113*, 140–156. [CrossRef]
18. Lee, W.L.; Gu, X.; Armas, F.; Chandra, F.; Chen, H.; Wu, F.; Leifels, M.; Xiao, A.; Chua, F.J.D.; Kwok, G.W.C.; et al. Quanti SARS-CoV-2 tracking of variants Delta, Delta plus, Kappa and Beta in wastewater by allele-specific RT-qPCR. *Environ. Sci. T Lett.* **2021**, *8*, 675–682. [CrossRef]
19. Vidic, J.; Vizzini, P.; Manzano, M.; Kavanaugh, D.; Ramarao, N.; Zivkovic, M.; Radonic, V.; Knezevic, N.; Giouroudi, I.; Gad I. Point-of-need DNA testing for detection of foodborne pathogenic bacteria. *Sensors* **2019**, *19*, 1100. [CrossRef]
20. Zhao, W.; Zhang, D.; Zhou, T.; Huang, J.; Wang, Y.; Li, B.; Chen, L.; Yang, J.; Liu, Y. Aptamer-conjugated magnetic Fe_3O core-shell multifunctional nanoprobe: A three-in-one aptasensor for selective capture, sensitive SERS detection and eff near-infrared light triggered photothermal therapy of *Staphylococcus aureus*. *Sens. Actuators B Chem.* **2022**, *350*, 130879. [Cro
21. Castillo-Henríquez, L.; Brenes-Acuña, M.; Castro-Rojas, A.; Cordero-Salmerón, R.; Lopretti-Correa, M.; Vega-Baudr Biosensors for the detection of bacterial and viral clinical pathogens. *Sensors* **2020**, *20*, 6926. [CrossRef]
22. Haleem, A.; Javaid, M.; Singh, R.P.; Suman, R.; Rab, S. Biosensors applications in medical field: A brief review. *Sens. Int.* *2*, 100100. [CrossRef]
23. Chadha, U.; Bhardwaj, P.; Agarwal, R.; Rawat, P.; Agarwal, R.; Gupta, I.; Panjwani, M.; Singh, S.; Ahuja, C.; Selvaraj, S.K. Recent progress and growth in biosensors technology: A critical review. *J. Ind. Eng. Chem.* **2022**, *109*, 21–51. [CrossRef]
24. Feng, X.; Li, P.; Li, T.; Cao, X.; Liu, D.; Xiao, M.; Wang, L. Ultra-sensitive and rapid detection of *Salmonella enterica* and *Staphylo aureus* to single-cell level by aptamer-functionalized carbon nanotube field-effect transistor biosensors. *Biosens. Bioelectron* *257*, 116333. [CrossRef]
25. Abedi, R.; Raoof, J.B.; Mohseni, M.; Hashkavayi, A.B. Sandwich-type electrochemical aptasensor based on hemin-graphite as a signal label and rGO/MWCNTs/chitosan/carbon quantum dot modified electrode for sensitive detection of *Acinetc baumannii* bacteria. *Anal. Chim. Acta* **2024**, *1303*, 342491. [CrossRef]
26. Tepeli, Y.; Anik, Ü. Electrochemical biosensors for influenza virus A detection: The potential of adaptation of these devi POC systems. *Sens. Actuators B Chem.* **2018**, *254*, 377–384. [CrossRef]
27. Li, D.; Liu, L.; Huang, Q.; Tong, T.; Zhou, Y.; Li, Z.; Bai, Q.; Liang, H.; Chen, L. Recent advances on aptamer-based biosens detection of pathogenic bacteria. *World J. Microbiol. Biotechnol.* **2021**, *37*, 45. [CrossRef]
28. Napit, R.; Jaysawal, S.K.; Chowdhury, R.; Catague, J.; Melke, H.; Pham, C.V.; Xu, H.; Jia, L.; Lin, J.; Hou, Y.; et al. Aptasenso advancement in molecular recognition technology. *Adv. Mater. Technol.* **2025**, *10*, 2400504. [CrossRef]
29. Sargazi, S.; Simge, E.R.; Mobashar, A.; Gelen, S.S.; Rahdar, A.; Ebrahimi, N.; Hosseinikhah, S.M.; Bilal, M.; Kyzas, G.Z. Ap conjugated carbon-based nanomaterials for cancer and bacteria theranostics: A review. *Chem. Biol. Interact.* **2022**, *361*, 1 [CrossRef] [PubMed]
30. Léguillier, V.; Heddi, B.; Vidic, J. Recent advances in aptamer-based biosensors for bacterial detection. *Biosensors* **2024**, *1* [CrossRef] [PubMed]
31. Lee, S.J.; Cho, J.; Lee, B.H.; Hwang, D.; Park, J.W. Design and prediction of aptamers assisted by in silico methods. *Biome* **2023**, *11*, 356. [CrossRef]
32. Ishida, R.; Adachi, T.; Yokota, A.; Yoshihara, H.; Aoki, K.; Nakamura, Y.; Hamada, M. RaptRanker: In silico RNA aptamer sel from HT-SELEX experiment based on local sequence and structure information. *Nucleic Acids Res.* **2020**, *48*, e82. [CrossRe

Cossettini, A.; Pasquardini, L.; Romani, A.; Feriani, A.; Pinamonti, D.; Manzano, M. Computational aptamer design for spike glycoprotein (S) (SARS CoV-2) detection with an electrochemical aptasensors. *Appl. Microbiol. Biotechnol.* **2024**, *108*, 259. [CrossRef]
Song, Y.; Song, J.; Wei, X.; Huang, M.; Sun, M.; Zhu, L.; Lin, B.; Shen, H.; Zhu, Z.; Yang, C. Discovery of aptamers targeting the receptor-binding domain of the SARS-CoV-2 spike glycoprotein. *Anal. Chem.* **2020**, *92*, 9895–9900. [CrossRef]
Abrego-Martinez, J.C.; Jafari, M.; Chergui, S.; Pavel, C.; Che, D.; Siaj, M. Aptamer-based electrochemical biosensor for rapid detection of SARS-CoV-2: Nanoscale electrode-aptamer-SARS-CoV-2 imaging by photo-induced force microscopy. *Biosens. Bioelectron.* **2022**, *195*, 113595. [CrossRef]
Wandtke, T.; Wędrowska, E.; Szczur, M.; Przybylski, G.; Libura, M.; Kopiński, P. Aptamers—Diagnostic and therapeutic solution in SARS-CoV-2. *Int. J. Mol. Sci.* **2022**, *23*, 1412. [CrossRef]
Sohouli, E.; Ghalkhani, M.; Zargar, T.; Joseph, Y.; Rahimi-Nasrabadi, M.; Ahmadi, F.; Plonska-Brzezinska, M.E.; Ehrlich, H. A new electrochemical aptasensor based on gold/nitrogen-doped carbon nano-onions for the detection of *Staphylococcus aureus*. *Electrochim. Acta* **2022**, *403*, 139633. [CrossRef]
Bypabekova, M.; Jolly, P.; Estrela, P.; Kanayeva, D. Electrochemical aptasensor using optimized surface chemistry for the detection of *Mycobacterium tuberculosis* secreted protein MPT64 in human serum. *Biosens. Bioelectron.* **2019**, *123*, 141–151. [CrossRef]
Banerjee, S.; Hemmat, M.A.; Shubham, S.; Gosai, A.; Devarakonda, S.; Jiang, N.; Geekiyanage, C.; Dillard, J.A.; Maury, W.; Shrotriya, P.; et al. Structurally different yet functionally similar: Aptamers specific for the Ebola virus soluble glycoprotein and GP1,2 and their application in electrochemical sensing. *Int. J. Mol. Sci.* **2023**, *24*, 4627. [CrossRef]
Suh, S.H.; Choi, S.J.; Dwivedi, H.P.; Moore, M.D.; Escudero-Abarca, B.I.; Jaykus, L.A. Use of DNA aptamer for sandwich type detection of *Listeria monocytogenes*. *Anal. Biochem.* **2018**, *557*, 27–33. [CrossRef]
Duan, N.; Chang, B.; Zhang, H.; Wang, Z.; Wu, S. *Salmonella typhimurium* detection using a surface-enhanced Raman scattering-based aptasensors. *Int. J. Food Microbiol.* **2016**, *218*, 38–43. [CrossRef]
Hamula, C.L.A.; Peng, H.; Wang, Z.; Tyrrell, G.J.; Li, X.-F.; Le, X.C. An improved SELEX technique for selection of DNA aptamers binding to M-type 11 of *Streptococcus pyogenes*. *Methods* **2016**, *97*, 51–57. [CrossRef]
Marton, S.; Cleto, F.; Krieger, M.A.; Cardoso, J. Isolation of an aptamer that binds specifically to *E. coli*. *PLoS ONE* **2016**, *11*, e0153637. [CrossRef] [PubMed]
Soundy, J.; Day, D. Selection of DNA aptamers specific for live *Pseudomonas aeruginosa*. *PLoS ONE* **2017**, *12*, e0185385. [CrossRef]
Babamiri, B.; Salimi, A.; Hallaj, R. A molecularly imprinted electrochemiluminescence sensor for ultrasensitive HIV-1 gene detection using EuS nanocrystals as luminophore. *Biosens. Bioelectron.* **2018**, *117*, 332–339. [CrossRef]
Li, X.; Yin, C.; Wu, Y.; Zhang, Z.; Jiang, D.; Xiao, D.; Fang, X.; Zhou, C. Plasmonic nanoplatform for point-of-care testing trace HCV core protein. *Biosens. Bioelectron.* **2020**, *147*, 111488. [CrossRef]
Chand, R.; Neethirajan, S. Microfluidic platform integrated with graphene-gold nano-composite aptasensor for one-step detection of norovirus. *Biosens. Bioelectron.* **2017**, *98*, 47–53. [CrossRef]
Lee, K.H.; Zeng, H. Aptamer-based ELISA assay for highly specific and sensitive detection of Zika NS1 protein. *Anal. Chem.* **2017**, *89*, 12743–12748. [CrossRef]
Chidzwondo, F.; Mutapi, F. Challenge of diagnosing acute infections in poor resource settings in Africa. *AAS Open Res. Afr.* **2024**, *4*, 28. [CrossRef]
Fàbrega, A.; Vila, J. *Salmonella enterica* serovar *Typhimurium* skills to succeed in the host: Virulence and regulation. *Clin. Microbiol.* **2013**, *26*, 308–341. [CrossRef] [PubMed]
Vallés, P.G.; Gil, L.A.F.; Cacciamani, V.; Benardon, M.E.; Costantino, V.V. Shiga toxin-producing *Escherichia coli* associated hemolytic uremic syndrome. *J. Urol. Nephrol.* **2023**, *8*, 000247. [CrossRef]
Jin, T.; Mohammad, M.; Pullerits, R.; Ali, A. Bacteria and host interplay in *Staphylococcus aureus* septic arthritis and sepsis. *Pathogens* **2021**, *10*, 158. [CrossRef]
Hmar, E.B.L.; Paul, S.; Sharma, H.K. The role of *Shigella spp.* in propagating bacillary dysentery in humans and the prominence of nanotechnology in disease prevention. *Future J. Pharm. Sci.* **2024**, *10*, 97. [CrossRef]
Wasik, D.; Mulchandani, A.; Yates, M.V. A heparin-functionalized carbon nanotube-based affinity biosensor for dengue virus. *Biosens. Bioelectron.* **2017**, *91*, 811–816. [CrossRef] [PubMed]
Mittler, E.; Dieterle, M.E.; Kleinfelter, L.M.; Slough, M.M.; Chandran, K.; Jangra, R.K. Hantavirus entry: Perspectives and recent advances. *Adv. Virus Res.* **2019**, *104*, 185–224. [CrossRef] [PubMed]
Kharsany, A.B.M.; McKinnon, L.R.; Lewis, L.; Cawood, C.; Khanyile, D.; Maseko, D.V.; Goodman, T.C.; Beckett, S.; Govender, K.; George, G.; et al. Population prevalence of sexually transmitted infections in a high HIV burden district in KwaZulu-Natal, South Africa: Implications for HIV epidemic control. *Int. J. Infect. Dis.* **2020**, *98*, 130–137. [CrossRef] [PubMed]
Rojas, M.; Monsalve, D.M.; Pacheco, Y.; Acosta-Ampudia, Y.; Ramírez-Santana, C.; Ansari, A.A.; Gershwin, M.E.; Anaya, J.M. Ebola virus disease: An emerging and re-emerging viral threat. *J. Autoimmun.* **2020**, *106*, 102375. [CrossRef]
Yin, T.-L.; Chen, N.; Zhang, J.-Y.; Yang, S.; Li, W.-M.; Gao, X.-H.; Shi, H.-L.; Hu, H.-P. Excess multi-cause mortality linked to Influenza virus infection in China, 2012–2021: A population-based study. *Front. Public Health* **2024**, *12*, 1399672. [CrossRef]

https://doi.org/10.3390/molecules31040664

59. Msemburi, W.; Karlinsky, A.; Knutson, V.; Aleshin-Guendel, S.; Chatterji, S.; Wakefield, J. The WHO estimates of excess mort associated with the COVID-19 pandemic. *Nature* **2023**, *613*, 130–137. [CrossRef]
60. Faramarzi, A.; Norouzi, S.; Dehdarirad, H.; Aghlmand, S.; Yusefzadeh, H.; Javan-Noughabi, J. The global economic burd COVID-19 disease: A comprehensive systematic review and meta-analysis. *Syst. Rev.* **2024**, *13*, 68. [CrossRef]
61. Scott, A.; Ansari, W.; Khan, F.; Chambers, R.; Benigno, M.; Di Fusco, M.; McGrath, L.; Malhotra, D.; Draica, F.; Nguyen, J.; Substantial health and economic burden of COVID-19 during the year after acute illness among US adults at high risk of se COVID-19. *BMC Med.* **2024**, *22*, 46. [CrossRef]
62. Bengoechea, J.A.; Bamford, C.G. SARS-CoV-2, bacterial co-infections, and AMR: The deadly trio in COVID-19. *EMBO Mol.* **2020**, *12*, e12560. [CrossRef] [PubMed]
63. Rawson, T.M.; Moore, L.S.P.; Zhu, N.; Ranganathan, N.; Skolimowska, K.; Gilchrist, M.; Satta, G.; Cooke, G.; Holmes, A. Bact and fungal coinfection in individuals with coronavirus: A rapid review to support COVID-19 antimicrobial prescribing. *Infect. Dis.* **2020**, *71*, 2459–2468. [CrossRef] [PubMed]
64. Hendaus, M.A.; Jomha, F.A. Covid-19 induced superimposed bacterial infection. *J. Biomol. Struct. Dyn.* **2020**, *39*, 4185– [CrossRef]
65. Almand, E.A.; Moore, M.D.; Jaykus, L.A. Virus-bacteria interactions: An emerging topic in human infection. *Viruses* **2017**, [CrossRef] [PubMed]
66. Mirzaei, R.; Goodarzi, P.; Asadi, M.; Soltani, A.; Aljanabi, H.A.A.; Jeda, A.S.; Dashtbin, S.; Jalalifar, S.; Mohammadzadeh Teimoori, A.; et al. Bacterial co-infections with SARS-CoV-2. *IUBMB Life* **2020**, *72*, 2097–2111. [CrossRef]
67. Thornhill, J.P.; Barkati, S.; Walmsley, S.; Rockstroh, J.; Antinori, A.; Harrison, L.B.; Palich, R.; Nori, A.; Reeves, I.; Habibi, et al. Monkeypox virus infection in humans across 16 Countries—April-June 2022. *N. Engl. J. Med.* **2022**, *387*, 679–691. [Cross
68. Marburg Virus Disease. Available online: https://www.who.int/news-room/fact-sheets/detail/marburg-virus-disease (acce on 30 July 2025).
69. Marburg Virus Disease—Ghana. Available online: https://www.who.int/emergencies/disease-outbreak-news/item/ don402 (accessed on 30 July 2025).
70. Zhao, F.; He, Y.; Lu, H. Marburg virus disease: A deadly rare virus is coming. *Biosci. Trends* **2022**, *16*, 312–316. [CrossRef]
71. Ebola Virus Disease. Available online: https://www.who.int/news-room/fact-sheets/detail/ebola-disease (accessed c July 2025).
72. Keita, A.K.; Koundouno, F.R.; Faye, M.; Düx, A.; Hinzmann, J.; Diallo, H.; Ayouba, A.; Marcis, F.L.; Soropogui, B.; Ifono, K.; Resurgence of Ebola virus in 2021 in Guinea suggests a new paradigm for outbreaks. *Nature* **2021**, *597*, 539–543. [CrossRef]
73. Xiong, W.M.; Xu, Q.P.; Li, X.; Xiao, R.D.; Cai, L.; He, F. The association between human papillomavirus infection and lung ca A system review and meta-analysis. *Oncotarget* **2017**, *8*, 96419–96432. [CrossRef]
74. Yang, X.; Qi, S.; Dai, L.; Ye, Q.; Li, X. Trends in HPV-positive cervical cancer prevalence: A retrospective study from 2013 to *Virol. J.* **2025**, *22*, 199. [CrossRef]
75. Zhang, C.-H.; Cheng, Y.; Zhang, S.; Fan, J.; Gao, Q. Changing epidemiology of hepatocellular carcinoma in Asia. *Liver Int.* *42*, 2029–2041. [CrossRef]
76. Sung, H.; Ferlay, J.; Siegel, R.L.; Laversanne, M.; Soerjomataram, I.; Jemal, A.; Bray, F. Global Cancer Statistics 2020: GLOBOC estimates of incidence and mortality worldwide for 36 cancers in 185 countries. *CA Cancer J. Clin.* **2021**, *71*, 209–249. [Cross
77. Martins, I.; Ribeiro, I.P.; Jorge, J.; Gonçalves, A.C.; Gonçalves, A.C.; Melo, J.B.; Carreira, I.M. Liquid biopsies: Application cancer diagnosis and monitoring. *Genes* **2021**, *12*, 349. [CrossRef] [PubMed]
78. Li, F.; Yu, Z.; Han, X.; Lai, R.Y. Electrochemical aptamer-based sensors for food and water analysis: A review. *Anal. Chim.* **2019**, *1051*, 1–23. [CrossRef] [PubMed]
79. Chen, X.; Gao, D.; Sun, F.; Li, Z.; Wang, Y.; Qiu, C.; He, K.; Wang, J. Nanomaterial-based aptamer biosensors for ochratox detection: A review. *Anal. Bioanal. Chem.* **2022**, *414*, 2953–2969. [CrossRef]
80. Choi, H.K.; Lee, J.; Park, M.; Oh, J.H. Development of single-walled carbon nanotube-based biosensor for the detecti *Staphylococcus aureus. J. Food Qual.* **2017**, *1*, 5239487. [CrossRef]
81. Fatemi, K.; Lau, S.Y.; Obayomi, K.S.; Kiew, S.F.; Coorey, R.; Chung, L.Y.; Fatemi, R.; Heshmatipour, Z.; Premarathna, K Carbon nanomaterial-based aptasensors for rapid detection of foodborne pathogenic bacteria. *Anal. Biochem.* **2024**, *695*, 11 [CrossRef]
82. Jalalvand, A.R.; Karami, M.M. Roles of nanotechnology in electrochemical sensors for medical diagnostic purposes: A re *Sens. Bio Sens. Res.* **2025**, *47*, 100733. [CrossRef]
83. Bisht, N.; Patel, M.; Dwivedi, N.; Kumar, P.; Mondal, D.P.; Srivastava, A.K.; Dhand, C. Bio-inspired polynorepinephrine b nanocoatings for reduced graphene oxide/gold nanoparticles composite for high-performance biosensing of *Mycobacte tuberculosis. Environ. Res.* **2023**, *227*, 115684. [CrossRef]
84. Gómez-López, P.; Puente-Santiago, A.; Castro-Beltrán, A.; do Nascimento, L.A.S.; Balu, A.M.; Luque, R.; Alvarado-Beltrán, Nanomaterials and catalysis for green chemistry. *Curr. Opin. Green Sustain. Chem.* **2020**, *24*, 48–55. [CrossRef]

Hammami, I.; Alabdallah, N.M.; Al Jomaa, A.; Kamoun, M. Gold nanoparticles: Synthesis properties and applications. *J. King Saud Univ. Sci.* **2021**, *33*, 101560. [CrossRef]

Kalimuthu, K.; Cha, B.S.; Kim, S.; Park, K.S. Eco-friendly synthesis and biomedical applications of gold nanoparticles: A review. *Microchem. J.* **2020**, *152*, 104296. [CrossRef]

Fahim, M.; Shahzaib, A.; Nishat, N.; Jahan, A.; Bhat, T.A.; Inam, A. Green synthesis of silver nanoparticles: A comprehensive review of methods, influencing factors, and applications. *JCIS Open* **2024**, *16*, 100125. [CrossRef]

Philip, D. Green synthesis of gold and silver nanoparticles using *Hibiscus rosa sinensis*. *Phys. E Low Dimens. Syst. Nanostruct.* **2010**, *42*, 1417–1424. [CrossRef]

Das, R.; Dhiman, A.; Kapil, A.; Bansal, V.; Sharma, T.K. Aptamer-mediated colorimetric and electrochemical detection of Pseudomonas aeruginosa utilizing peroxidase-mimic activity of gold NanoZyme. *Anal. Bioanal. Chem.* **2019**, *411*, 1229–1238. [CrossRef]

Zarei, S.S.; Soleimanian-Zad, S.; Ensafi, A.A. An impedimetric aptasensor for *Shigella dysenteriae* using a gold nanoparticle-modified glassy carbon electrode. *Microchim. Acta* **2018**, *185*, 538. [CrossRef] [PubMed]

Ge, C.; Yuan, R.; Yi, L.; Yang, J.; Zhang, H.; Li, L.; Nian, W.; Yi, G. Target-induced aptamer displacement on gold nanoparticles and rolling circle amplification for ultrasensitive live *Salmonella typhimurium* electrochemical biosensing. *J. Electroanal. Chem.* **2018**, *826*, 174–180. [CrossRef]

Lee, T.; Park, S.Y.; Jang, H.; Kim, G.-H.; Lee, Y.; Park, C.; Mohammadniaei, M.; Lee, M.-H.; Min, J. Fabrication of electrochemical biosensor consisted of multi-functional DNA structure/porous Au nanoparticle for avian influenza virus (H5N1) in chicken serum. *Mater. Sci. Eng. C* **2019**, *99*, 511–519. [CrossRef]

Khan, R.; Deshpande, A.S.; Proteasa, G.; Andreescu, S. Aptamer-based electrochemical biosensor with S protein binding affinity for COVID-19 detection: Integrating computational design with experimental validation of S protein binding affinity. *Sens. Actuators B Chem.* **2024**, *399*, 134775. [CrossRef]

Cajigas, S.; Alzate, D.; Orozco, J. Gold nanoparticle/DNA-based nanobioconjugate for electrochemical detection of Zika virus. *Microchim. Acta* **2020**, *187*, 594. [CrossRef]

Rhouati, A.; Zourob, M. Development of a multiplexed electrochemical aptasensor for the detection of cyanotoxins. *Biosensors* **2024**, *14*, 268. [CrossRef]

Roushani, M.; Sarabaegi, M.; Pourahmad, F. Impedimetric aptasensor for *Pseudomonas aeruginosa* by using a glassy carbon electrode modified with silver nanoparticles. *Microchim. Acta* **2019**, *186*, 725. [CrossRef]

Abbaspour, A.; Norouz-Sarvestani, F.; Noori, A.; Soltani, N. Aptamer-conjugated silver nanoparticles for electrochemical dual-aptamer-based sandwich detection of *Staphylococcus aureus*. *Biosens. Bioelectron.* **2015**, *68*, 149–155. [CrossRef]

Zeng, W.; Tang, X.; Wu, T.; Han, B.; Wu, L. Development of a highly sensitive aptamer-based electrochemical sensor for detecting saxitoxin based on $K_3Fe(CN)_6$ regulated silver nanoparticles. *Anal. Chim. Acta* **2024**, *1287*, 342134. [CrossRef]

Wang, L.; Huo, X.; Qi, W.; Xia, Z.; Li, Y.; Lin, J. Rapid and sensitive detection of *Salmonella typhimurium* using nickel nanowire bridge for electrochemical impedance amplification. *Talanta* **2020**, *211*, 120715. [CrossRef]

Jamal, R.B.; Gosewinkel, U.B.; Ferapontova, E.E. Electrocatalytic aptasensor for bacterial detection exploiting ferricyanide reduction by methylene blue on mixed PEG/aptamer monolayers. *Bioelectrochemistry* **2024**, *156*, 108620. [CrossRef]

Li, N.; Huang, X.; Sun, D.; Yu, W.; Tan, W.; Luo, Z.; Chen, Z. Dual-aptamer-based voltammetric biosensor for the *Mycobacterium tuberculosis* antigen MPT64 by using a gold electrode modified with a peroxidase loaded composite consisting of gold nanoparticles and a Zr(IV)/terephthalate metal-organic framework. *Microchim. Acta* **2018**, *185*, 543. [CrossRef]

Cho, I.H.; Kim, D.H.; Park, S. Electrochemical biosensors: Perspective on functional nanomaterials for on-site analysis. *Biomat. Res.* **2020**, *24*, 6. [CrossRef] [PubMed]

Gao, H.; Xu, T.; Zhou, J.; Rojas, O.; He, M.; Ji, X.; Dai, H. Electrochemical sensing of *Staphylococcus aureus* based on conductive anti-fouling interface. *Microchim. Acta* **2022**, *189*, 97. [CrossRef] [PubMed]

Ranjbar, S.; Shahrokhian, S. Design and fabrication of an electrochemical aptasensor using Au nanoparticles/carbon nanoparticles/cellulose nanofibers nanocomposite for rapid and sensitive detection of *Staphylococcus aureus*. *Bioelectrochemistry* **2018**, *123*, 70–76. [CrossRef] [PubMed]

Bai, L.; Chen, Y.; Bai, Y.; Chen, Y.; Zhou, J.; Huang, A. Fullerene-doped polyaniline as new redox nanoprobe and catalyst in electrochemical aptasensor for ultrasensitive detection of *Mycobacterium tuberculosis* MPT64 antigen in human serum. *Biomaterials* **2017**, *133*, 11–19. [CrossRef]

Jaanei, M.; Taheri, R.A.; Eskandari, K. Electrochemical aptasensor for *Escherichia coli* O157:H7 bacteria detection using a nanocomposite of reduced graphene oxide, gold nanoparticles and polyvinyl alcohol. *Anal. Methods* **2021**, *13*, 3101–3109. [CrossRef]

Rafique, S.; Akram, R.; Nasir, R.; Naz, N.; Shafique, A.; Bashir, S.; Haq, Z. An aptamer and hydroxyapatite-silver-zinc oxide–based novel electrochemical sensor for ultrasensitive *H. pylori detection*. *Microchim. Acta* **2024**, *191*, 751. [CrossRef]

https://doi.org/10.3390/molecules31040664

108. Jiang, X.; Lv, Z.; Rao, C.; Chen, X.; Zhang, Y.; Lin, F. Simple and highly sensitive electrochemical detection of *Listeria monocyt*
based on aptamer-regulated Pt nanoparticles/hollow carbon spheres nanozyme activity. *Sens. Actuators B Chem.* **2023**, *392*, 1
[CrossRef]
109. Jiang, H.; Sun, Z.; Zhang, C.; Weng, X. 3D-architectured aptasensor for ultrasensitive electrochemical detection of norovirus
on phosphorene-gold nanocomposites. *Sens. Actuators B Chem.* **2022**, *354*, 131232. [CrossRef]
110. Zhong, H.; Yu, C.; Gao, R.; Chen, J.; Yu, Y.; Geng, Y.; Wen, Y.; He, J. A novel sandwich aptasensor for detecting T-2 toxin bas
rGO-TEPA-Au@Pt nanorods with a dual signal amplification strategy. *Biosens. Bioelectron.* **2019**, *144*, 111635. [CrossRef]
111. Nodoushan, S.M.; Nasirizadeh, N.; Sedighian, H.; Kachuei, R.; Azimzadeh-Taft, M.; Fooladi, A.A.I. Detection of Staphylo
Enterotoxin A (SEA) using a sensitive nanomaterial-based electrochemical aptasensor. *Diam. Relat. Mater.* **2022**, *127*, 1
[CrossRef]
112. Nikolova, M.P.; Chavali, M.S. Metal oxide nanoparticles as biomedical materials. *Biomimetics* **2020**, *5*, 27. [CrossRef] [Pub
113. Zhang, H.; Yao, S.; Song, X.; Xu, K.; Wang, J.; Li, J.; Zhao, C.; Jin, M. One-step colorimetric detection of *Staphylococcus a*
based on target-induced shielding against the peroxidase mimicking activity of aptamer-functionalized gold-coated iron
nanocomposites. *Talanta* **2021**, *232*, 122448. [CrossRef]
114. Jiang, D.; Ni, D.; Rosenkrans, Z.T.; Huang, P.; Yan, X.; Cai, W. Nanozyme: New horizons for responsive biomedical applic
Chem. Soc. Rev. **2019**, *48*, 3683–3704. [CrossRef]
115. Sharma, P.; Hassan, H.; Hasan, M.R.; Fatima, T.; Mosina, K.M.; Shukla, S.K.; Narang, J. Enhanced electrochemical det
of chikungunya virus through aptasensor modification with origami paper-based ternary nanocomposite. *Microchem. J.*
203, 110894. [CrossRef]
116. El-Wekil, M.M.; Halby, H.M.; Darweesh, M.; Ali, M.E.; Ali, R. An innovative dual recognition aptasensor for specific detect
Staphylococcus aureus based on Au/Fe_3O_4 binary hybrid. *Sci. Rep.* **2022**, *12*, 12502. [CrossRef]
117. Ming, P.; Lai, H.; Liu, Y.; Wang, J.; You, F.; Sun, D.; Zhai, H. Aptasensor development for T-2 toxin detection utilizing a
signal amplification strategy: Synergistic effects of bimetallic oxide (Ce-In)Ox and COFTAPB-DMTP. *Sens. Actuators B Chem.*
396, 134602. [CrossRef]
118. Zhang, S.; Wang, Y.; Sheng, Q.; Yue, T. Electrochemical aptasensor based on ZnO-Au nanocomposites for the determinat
ochratoxin A in wine and beer. *Processes* **2023**, *11*, 864. [CrossRef]
119. Evtugyn, G.; Porfireva, A.; Shamagsumova, R.; Hianik, T. Advances in electrochemical aptasensors based on carbon nanoma
Chemosensors **2020**, *8*, 96. [CrossRef]
120. Ayodele, O.O.; Adesina, A.O.; Pourianejad, S.; Averitt, J.; Ignatova, T. Recent advances in nanomaterial-based aptasens
medical diagnosis and therapy. *Nanomaterials* **2021**, *11*, 932. [CrossRef] [PubMed]
121. Kaur, H.; Shorie, M.; Sharma, M.; Ganguli, A.K.; Sabherwal, P. Bridged rebar graphene functionalized aptasensor for path
E. coli O78:K80:H11 detection. *Biosens. Bioelectron.* **2017**, *98*, 486–493. [CrossRef]
122. Hasan, M.R.; Pulingam, T.; Appaturi, J.N.; Zifruddin, A.N.; Teh, S.J.; Lim, T.W.; Ibrahim, F.; Leo, B.F.; Thong, K.L. C
nanotube-based aptasensor for sensitive electrochemical detection of whole-cell *Salmonella*. *Anal. Biochem.* **2018**, *554*,
[CrossRef]
123. Pathania, P.K.; Saini, J.K.; Vij, S.; Tewari, R.; Sabherwal, P.; Rishi, P.; Suri, C.R. Aptamer functionalized MoS_2-rGO nanocom
based biosensor for the detection of Vi antigen. *Biosens. Bioelectron.* **2018**, *122*, 121–126. [CrossRef]
124. Li, L.; Yuan, Y.; Chen, Y.; Zhang, P.; Bai, Y.; Bai, L. Aptamer based voltammetric biosensor for *Mycobacterium tuberculosis* a
ESAT-6 using a nanohybrid material composed of reduced graphene oxide and a metal-organic framework. *Microchim. Act*
185, 379. [CrossRef]
125. Muniandy, S.; Dinshaw, I.J.; Teh, S.J.; Lai, C.W.; Ibrahim, F.; Thong, K.L.; Leo, B.F. Graphene-based label-free electroche
aptasensor for rapid and sensitive detection of foodborne pathogen. *Anal. Bioanal. Chem.* **2017**, *409*, 6893–6905. [CrossRef]
126. Dinshaw, I.J.; Muniandy, S.; Teh, S.J.; Ibrahim, F.; Leo, B.F.; Thong, K.L. Development of an aptasensor using reduced gra
oxide chitosan complex to detect *Salmonella*. *J. Electroanal. Chem.* **2017**, *806*, 88–96. [CrossRef]
127. Muniandy, S.; Teh, S.J.; Appaturi, J.N.; Thong, K.L.; Lai, C.W.; Ibrahim, F.; Leo, B.F. A reduced graphene oxide-titanium d
nanocomposite based electrochemical aptasensor for rapid and sensitive detection of *Salmonella enterica*. *Bioelectrochemistr*
127, 136–144. [CrossRef]
128. Appaturi, J.N.; Pulingam, T.; Thong, K.L.; Muniandy, S.; Ahmad, N.; Leo, B.F. Rapid and sensitive detection of *Salmonell*
reduced graphene oxide-carbon nanotube based electrochemical aptasensors. *Anal. Biochem.* **2020**, *589*, 113489. [Cro
[PubMed]
129. Pourmadadi, M.; Shayeh, J.S.; Omidi, M.; Yazdian, F.; Alebouyeh, M.; Tayebi, L. A glassy carbon electrode modified with re
graphene oxide and gold nanoparticles for electrochemical aptasensing of lipopolysaccharides from *Escherichia coli* ba
Microchim. Acta **2019**, *186*, 787. [CrossRef]

Zamani, M.; Pourmadadi, M.; Seyyed Ebrahimi, S.A.; Yazdian, F.; Shabani Shayeh, J. A novel labeled and label-free dual electrochemical detection of endotoxin based on aptamer-conjugated magnetic reduced graphene oxide-gold nanocomposite. *J. Electroanal. Chem.* **2022**, *908*, 116116. [CrossRef]
Mohsin, D.H.; Mashkour, M.S.; Fatemi, F. Design of aptamer-based sensing platform using gold nanoparticles functionalized reduced graphene oxide for ultrasensitive detection of Hepatitis B virus. *Chem. Pap.* **2021**, *75*, 279–295. [CrossRef]
Ghanbari, K.; Roushani, M. A nanohybrid probe based on double recognition of an aptamer MIP grafted onto a MWCNTs-Chit nanocomposite for sensing hepatitis C virus core antigen. *Sens. Actuators B Chem.* **2018**, *258*, 1066–1071. [CrossRef]
Rahmati, Z.; Roushani, M.; Hosseini, H. Three-dimensional $NiCo_2O_4$ nanowires encapsulated in nitrogen-doped carbon networks as a high-performance aptamer stabilizer for impedimetric ultrasensitive detection of hepatitis C virus core antigen. *Surf. Interf* **2021**, *22*, 100813. [CrossRef]
Chekin, F.; Bagga, K.; Subramanian, P.; Jijie, R.; Singh, S.K.; Kurungot, S.; Boukherroub, R.; Szunerits, S. Nucleic aptamer modified porous reduced graphene oxide/MoS_2 based electrodes for viral detection: Application to human papillomavirus (HPV). *Sens. Actuators B Chem.* **2018**, *262*, 991–1000. [CrossRef]
Bao, W.; Aodeng, G.; Ga, L.; Ai, J. Aptamer-based electrochemical biosensors: Signal transduction mechanisms, application progress, and future trends. *Sens. Actuator Rep.* **2025**, *10*, 100366. [CrossRef]
Abd-Ellatief, R.; Abd-Ellatief, M.R. Electrochemical aptasensors: Current status and future perspectives. *Diagnostics* **2021**, *11*, 104. [CrossRef]
Khan, I.; Saeed, K.; Khan, I. Nanoparticles: Properties, applications and toxicities. *Arab. J. Chem.* **2019**, *12*, 908–931. [CrossRef]
Oberhaus, F.V.; Frense, D.; Beckmann, D. Immobilization techniques for aptamers on gold electrodes for the electrochemical detection of proteins: A review. *Biosensors* **2020**, *10*, 45. [CrossRef]
Carroll, D.P.; Mendes, P.M. Recent advances in surface modification and antifouling strategies for electrochemical sensing in complex biofluids. *Curr. Opin. Electrochem.* **2023**, *40*, 101319. [CrossRef]
Song, Z.; Han, R.; Yu, K.; Li, R.; Luo, X. Antifouling strategies for electrochemical sensing in complex biological media. *Microchim. Acta* **2024**, *191*, 3. [CrossRef] [PubMed]
Noguchi, T.; Nishitani, S.; Sakata, T. Anti-biofouling effect of nanoporous gold electrode on nonspecific signal reduction for electrochemical biosensors. *J. Electrochem. Soc.* **2024**, *171*, 077507. [CrossRef]
Jiang, H.; Sun, Z.; Guo, Q.; Weng, X. Microfluidic thread-based electrochemical aptasensor for rapid detection of *Vibrio parahaemolyticus*. *Biosens. Bioelectron.* **2021**, *182*, 113191. [CrossRef]

https://doi.org/10.3390/molecules31040664

molecules

Review

DNA Nanostructure-Assembled Metallic Nanoparticles for Biosensing Applications

Shaokang Ren [1,†], Kai He [1,†], Canlin Cui [1], Haoyu Fan [1], Hongzhen Peng [1], Kai Jiao [1,*] and Lihua Wang [1,2,*]

1 Institute of Materiobiology, College of Sciences, Shanghai University, Shanghai 200444, China; iamskren@163.com (S.R.); hekai2511@163.com (K.H.); cuicanlin@shu.edu.cn (C.C.); nana7mi@shu.edu.cn (H.F.); penghongzhen@shu.edu.cn (H.P.)
2 Shanghai Collaborative Innovation Center of Intelligent Sensing Chip Technology, Shanghai University, Shanghai 200444, China
* Correspondence: kjiao@shu.edu.cn (K.J.); wanglihua@shu.edu.cn (L.W.)
† These authors contributed equally to this work.

Abstract

DNA nanotechnology offers an unprecedented level of structural programmability for organizing metallic nanoparticles into precisely defined architectures, providing a powerful platform for plasmonic biosensing. In particular, gold and silver nanoparticles assembled on DNA nanostructures enable nanometer-scale control over interparticle distance, orientation, and spatial symmetry, which directly govern collective plasmonic behaviors and optical signal transduction. This review summarizes recent advances in DNA nanostructure-mediated assembly of metal nanoparticles, with an emphasis on design principles and assembly strategies that enable static and dynamic control of nanoparticle organization. Representative examples are discussed to illustrate how well-defined plasmonic assemblies give rise to tunable optical responses, including localized surface plasmon resonance modulation, chiroptical signals, fluorescence enhancement or quenching, and surface-enhanced Raman scattering. The role of structural programmability and stimulus-responsive reconfiguration in translating molecular recognition events into amplified optical outputs is highlighted in the context of biosensing. Finally, current challenges and future perspectives are outlined, focusing on structural robustness, signal reproducibility, and integration toward practical and multiplexed biosensing platforms.

Keywords: DNA nanostructures; metal nanoparticles; plasmonic properties; biosensing

1. Introduction

The development of highly sensitive, specific, and multiplexed biosensors has become increasingly important for biomedical diagnostics, environmental monitoring, and point-of-care applications. Achieving high sensor performance requires not only accurate molecular recognition but also precise control over signal transduction processes [1,2]. Metallic nanoparticles, particularly gold and silver nanoparticles, have emerged as key functional nanomaterials due to their unique plasmonic properties. Their size- and shape-dependent optical responses enable enhanced sensitivity, tunable signal outputs, and versatile detection modalities, making them widely employed in biosensing platforms [3,4].

To fully exploit these properties, metallic nanoparticles must be organized into well-defined architectures. The spatial positions, orientations, and interparticle distances critically influence collective plasmonic effects, electromagnetic field enhancement, and electron transfer pathways, all of which directly impact signal amplification and detection sensitivity [5–7]. Well-controlled assemblies therefore offer improved reproducibility, enhanced signal output, and opportunities for integrating multiple sensing functions, which are difficult to achieve with randomly distributed nanoparticles.

However, conventional nanoparticle assembly strategies often suffer from limited spatial precision, structural heterogeneity, and poor batch-to-batch consistency [8–10]. Random aggregation or nonspecific adsorption can lead to heterogeneous structures, uncontrolled interparticle distances, and variable signal outputs, which significantly hinder sensor performance. Overcoming these challenges requires scaffolds capable of directing nanoparticle assembly with nanometer-scale accuracy while offering programmability and dynamic responsiveness.

DNA nanotechnology provides a versatile solution to these challenges [11,12]. Through the rational design of DNA tiles [13–17], origami [18–22], and wireframe structures [23,24], nanoparticles can be positioned at defined locations with sub-10 nm precision. Importantly, DNA nanostructures function not only as passive templates but also as active components that regulate nanoparticle spatial organization, plasmonic coupling, and signal transduction, functioning as integral components of functional biosensors. This approach enables programmable, reproducible, and multifunctional assemblies that can be dynamically reconfigured in response to specific molecular inputs.

While numerous excellent reviews have summarized biosensors based on DNA-functionalized metallic nanoparticles [25,26], this review focuses specifically on gold and silver nanoparticles assembled through DNA nanostructures, emphasizing design principles and assembly strategies, with representative examples illustrating signal transduction mechanisms and biosensing applications. We also discuss current challenges and future directions for integrating DNA nanostructures with gold and silver nanoparticles to construct high-performance, programmable biosensors.

2. DNA Nanostructures as Programmable Scaffolds

DNA nanotechnology provides an unprecedented platform for constructing highly precise and programmable nanoscale architectures, enabling the controlled organization of metallic nanoparticles [27,28]. DNA origami, in particular, has become a cornerstone technique in this field, offering a versatile and robust scaffold for the assembly of complex nanostructures. Initially introduced by Rothemund in 2006, the scaffolded DNA origami technique uses a long single-stranded DNA scaffold, which is folded into well-defined two-dimensional (2D) shapes through hybridization with hundreds of short staple strands in a one-pot annealing reaction [18]. This method enables the fabrication of planar structures up to ~100 nm in size with ~6 nm resolution, facilitating the creation of a wide variety of shapes, such as rectangles, stars, smiley faces, and triangles (Figure 1a). The simplicity, low cost, and modularity of this approach quickly spurred more complex designs, including asymmetric forms like a map of China [29], curved concentric rings [22], and even a Möbius strip [30].

Building on this 2D foundation, DNA nanostructures have been extended into three dimensions through techniques such as multi-layer stacking of DNA helices. Notable examples include the work by Douglas et al., who created a range of three-dimensional (3D) objects such as monoliths, square nuts, railed bridges, genie bottles, stacked crosses, and slotted crosses using a honeycomb-pleated lattice, with tunable dimensions between 10–100 nm (Figure 1b) [21]. Subsequently, Dietz et al. advanced the design by introducing precise, programmable bending and twisting of 3D nanostructures via base-pair insertions and deletions, resulting in intricate shapes like wireframe beach balls, gears, and spirals (Figure 1c) [20]. Additionally, Han et al. demonstrated a curvature-centric approach

https://doi.org/10.3390/molecules31030513

to designing 3D structures, coupling longitudinal and latitudinal bending to assemble smoothly curved geometries such as hemispheres, spheres, and nanoflasks, thus expanding the structural repertoire beyond angular, raster-based designs (Figure 1d) [22].

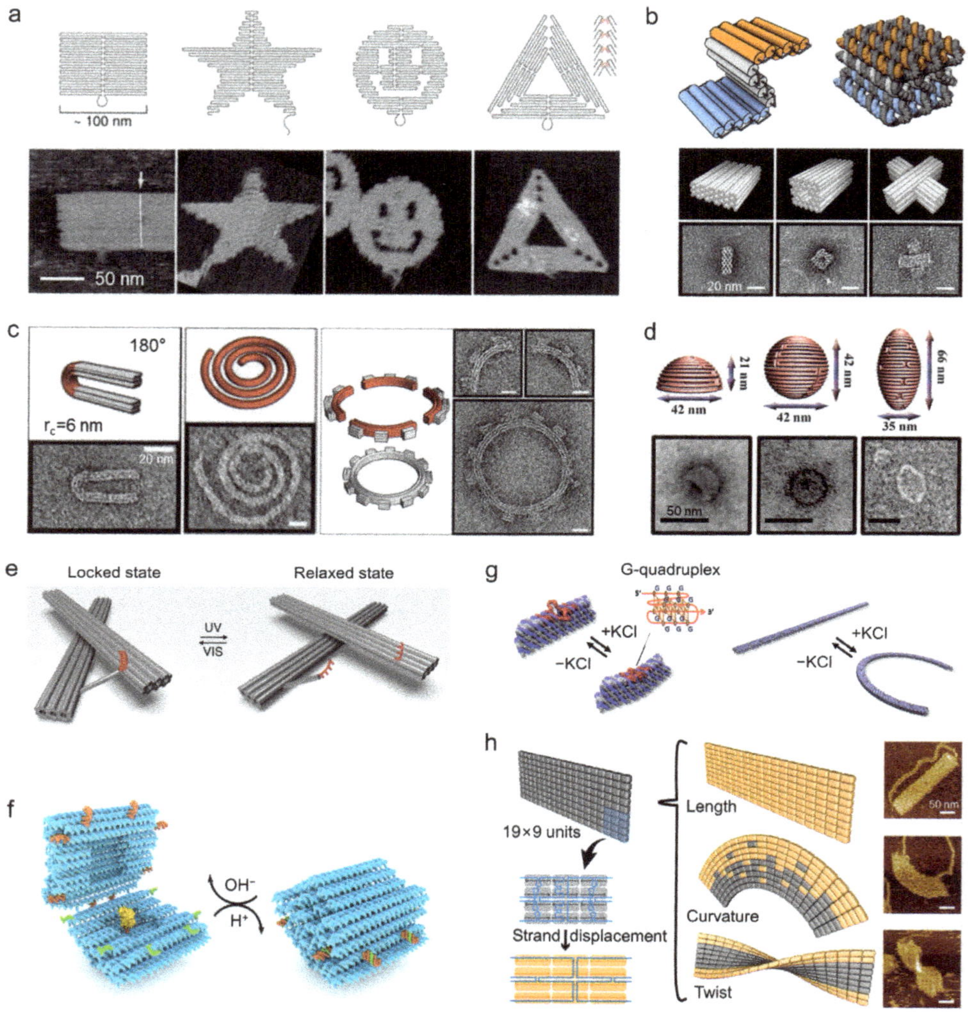

Figure 1. DNA origami nanostructures. (**a**) Symmetric 2D monolayer DNA origami structures: rectangle, five-pointed star, smiley face, and triangle (left to right) [31]. Copyright 2024, American Chemical Society. (**b**) 3D multilayer DNA origami structures with honeycomb packing [32]. Copyright 2016, Wiley. (**c**) 3D DNA origami structures exhibiting complex curvature [20]. Copyright 2009, AAAS. (**d**) DNA nanostructures with gradually varying radii, including hemispheres, spheres, and ellipsoids [22]. Copyright 2011, AAAS. (**e**) Conformational transitions of DNA origami mediated by light-responsive components [33]. Copyright 2016, Springer Nature. (**f**) Opening and closing of DNA nanocapsules regulated by pH-responsive modules. Green and orange represent the pH response module, while yellow represents the cargo inside the nanostructure [34]. Copyright 2019, American Chemical Society. (**g**) Global deformation of DNA origami via metal ion-responsive G-quadruplex motifs [35]. Copyright 2020, Wiley. (**h**) Tuning of length, curvature, and twist in DNA origami through toehold-mediated strand displacement [36]. Copyright 2021, American Chemical Society.

A significant leap in DNA nanotechnology has been the integration of dynamic and stimuli-responsive elements into otherwise static structures, transforming them into adaptive, responsive systems. For instance, Kuzyk et al. developed a light-driven reconfigurable DNA nanostructure by incorporating azobenzene-modified strands [33]. The photo-induced switching between trans and cis states triggered a macroscopic conformational change, which was amplified ~100-fold from the molecular length change (Figure 1e). Additionally, pH-responsive DNA structures have been realized by Ijäs et al., who designed a nanocapsule that opens at high pH for cargo loading and closes via Hoogsteen triplex formation at low pH, encapsulating gold nanoparticles in the process (Figure 1f) [34]. Ion-triggered reconfiguration was demonstrated by Suzuki, who constructed a nanoarm based on G-quadruplex-containing modules. This nanoarm contracts and relaxes reversibly upon the addition of K^+, toggling between linear and arched states (Figure 1g) [35].

Further enhancing the versatility of DNA scaffolds, modular reconfiguration has also been explored. Wang et al. introduced a modular expandable origami system wherein selective strand-displacement reactions allow for the independent adjustment of local length, curvature, and twist of a 2D sheet, enabling the controlled morphing of the shape (Figure 1h) [36]. This strategy demonstrates the potential for building more sophisticated, programmable DNA nanostructures that can dynamically adapt to specific environmental cues or functional requirements.

These advances in static and dynamic DNA nanostructures illustrate the profound potential of DNA nanotechnology as a programmable scaffold for organizing nanoparticles with high precision. By leveraging these DNA scaffolds, researchers can create nanoparticle assemblies that are not only spatially defined but also able to respond to external stimuli, enabling new opportunities in biosensing, where dynamic control over the structure and behavior of nanoparticle assemblies is crucial.

3. DNA Nanostructure-Mediated Assembly of Metal Nanoparticles

Building on the structural programmability of DNA nanostructures, gold and silver nanoparticles can be assembled into well-defined architectures with precise spatial control. At the most fundamental level, individual geometric parameters, such as interparticle distance and angular orientation [37], can be tuned to modulate plasmonic interactions. For example, silver nanoparticle dimers assembled on triangular DNA origami achieved center-to-center distances of ~90 nm, ~49 nm, and ~24 nm by adjusting the positions of capture strands (Figure 2a), illustrating nanometer-scale control over near-field coupling [38]. Angular orientation can also be precisely regulated, as shown by gold nanorod dimers on DNA origami templates, where 0°, 60°, 90°, and 180° configurations with gaps down to ~5 nm were realized (Figure 2b), enabling directional plasmonic responses [39].

Extending this precise control to multi-particle assemblies allows the creation of hierarchical arrays. One-dimensional arrangements of nanoparticles with uniform spacing form plasmonic waveguides capable of directing energy to distal reporters, exemplified by DNA-guided nanowire systems (Figure 2c) [40]. Two-dimensional lattices constructed with controlled interparticle spacing generate collective interference effects, such as Fano resonances, enhancing surface-enhanced Raman scattering (SERS) sensitivity (Figure 2d) [41]. Further, by positioning multiple DNA origami units in defined orientations on a surface, these patterned assemblies can form superlattices or metasurfaces, providing a structural basis for exploring nonlinear optical responses and collective field enhancement in macroscopic arrays (Figure 2e) [42].

Beyond planar configurations, three-dimensional assembly strategies expand the functional repertoire. By arranging nanoparticles into left- or right-handed helices, DNA scaffolds produce pronounced chiroptical responses, including circular dichroism (CD)

https://doi.org/10.3390/molecules31030513

signals (Figure 2f) [43]. Such 3D architectures not only increase structural diversity but also directly contribute to biosensing, as molecular recognition events can induce measurable changes in the chiral configuration.

Collectively, these hierarchical DNA-mediated assembly strategies demonstrate the versatility of DNA nanostructures in guiding nanoparticles from precise control of interparticle distance and angle, through one- and two-dimensional patterning, to three-dimensional helical architectures. This programmability enables the construction of plasmonic waveguides, Raman-enhancing lattices, and metasurfaces with potential for nonlinear optical applications, while simultaneously supporting sensitive, tunable, and multifunctional biosensing platforms.

While static DNA origami architectures provide precise and robust control over nanoparticle spacing and geometry, thereby maximizing plasmonic coupling and signal intensity, the incorporation of dynamic, stimuli-responsive elements enables these assemblies to actively transduce molecular inputs into measurable optical signals. In biosensing contexts, such reconfigurability is not merely structural but functional, as it directly translates molecular recognition events into signal amplification, reversibility, and background suppression.

A representative mechanism for dynamic signal transduction is toehold-mediated strand displacement, which enables programmable and reversible repositioning of nanoparticles on DNA scaffolds. Zhou et al. demonstrated an active plasmonic "walker" system in which a gold nanorod moved stepwise along a DNA track with ~7 nm spacing via strand displacement (Figure 2g) [45]. Importantly, the walker motion induced continuous and reversible changes in circular dichroism, illustrating how molecular inputs can be converted into real-time plasmonic signal outputs. Such systems exemplify how dynamic reconfiguration enables signal switching and temporal control, rather than static signal readout.

Light-responsive DNA origami further highlights how external stimuli can modulate plasmonic responses with high spatiotemporal precision. Kuzyk et al. developed a reconfigurable DNA origami template in which azobenzene-modified hinges allowed reversible switching between locked and relaxed states under UV/Vis illumination [33]. This molecular-scale actuation was amplified into a ~30 nm structural displacement, resulting in pronounced changes in chiral plasmonic responses in gold nanorod assemblies (Figure 2h). The ability to reversibly toggle plasmonic signals without chemical reagents offers clear advantages for background suppression and repeated sensing cycles.

Chemical stimuli such as pH provide an additional route for dynamic plasmonic modulation. By replacing azobenzene hinges with pH-sensitive DNA locks, Kuzyk et al. achieved reversible switching of plasmonic chirality in DNA-based metamolecules (Figure 2i) [46]. Similarly, Julin et al. designed pH-responsive DNA origami units whose opening and closing triggered large-scale reorganization of nanoparticle arrays, altering interparticle distances and lattice geometry (Figure 2j) [47]. In both cases, environmental changes were directly translated into distinct plasmonic states, enabling signal gating and condition-dependent readout.

Collectively, these examples illustrate how static DNA origami structures function as plasmonic signal amplifiers by defining optimal nanoparticle arrangements, whereas dynamic DNA nanotechnology enables signal transduction by converting molecular or environmental stimuli into reversible optical responses. The integration of static amplification with dynamic reconfiguration thus provides a powerful framework for biosensing, allowing high sensitivity, low background, and real-time adaptability within a single DNA-programmed plasmonic platform.

https://doi.org/10.3390/molecules310

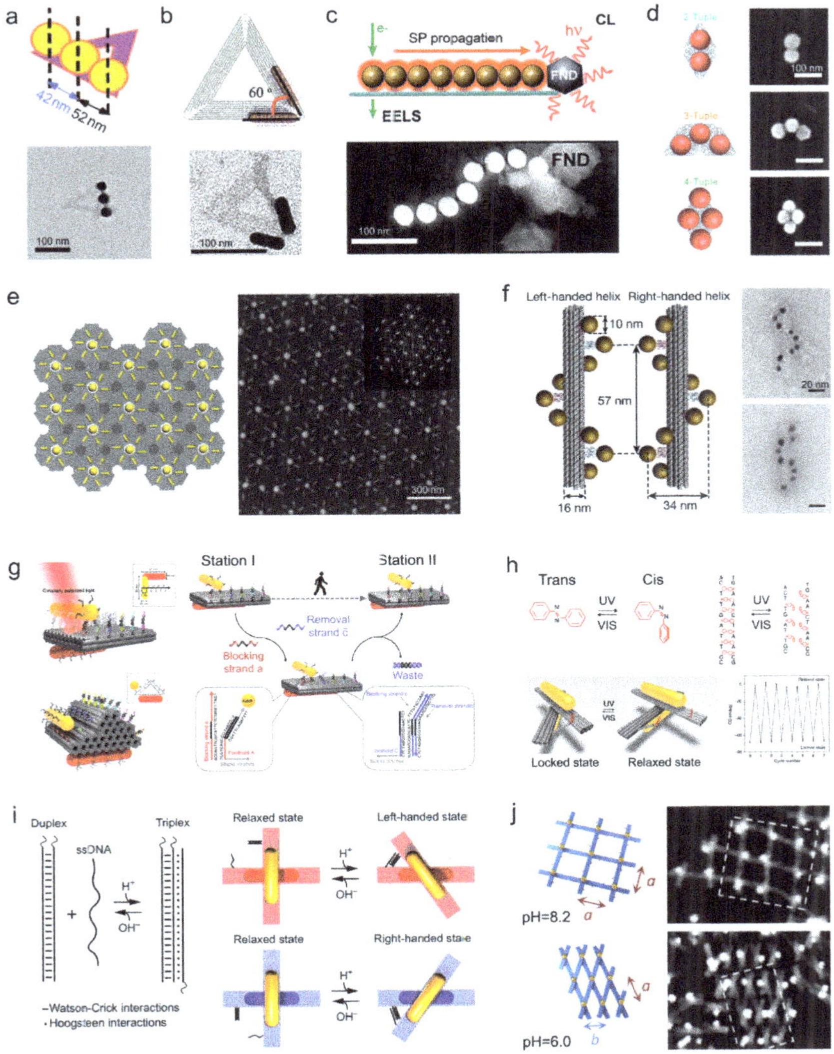

Figure 2. DNA Nanostructure-Mediated MNP architectures. (**a**) Precise control over the position, number, and spacing of AgNPs using triangular DNA origami [38]. Copyright 2010, Wiley. (**b**) Angle regulation of AuNRs based on triangular DNA origami templates [39]. Copyright 2011, American Chemical Society. (**c**) DNA-origami-based self-assembly of gold nanoparticle chains for plasmonic waveguides [40]. Copyright 2018, American Chemical Society. (**d**) AuNP metamolecules based on super-origami templates [41]. Copyright 2019, AAAS. (**e**) Fabrication of chiral metasurface via microscale DNA origami array [42]. Copyright 2025, American Chemical Society. (**f**) Construction of helical AuNP assemblies with optical chirality responses utilizing DNA origami 24-helix bundles [44]. Copyright 2018, American Chemical Society. (**g**) A DNA origami-based AuNR walker system driven by toehold-mediated DNA strand displacement reactions [45]. Copyright 2015, Springer Nature. (**h**) A light-responsive DNA origami template to construct a plasmonic chiral nanosystem [33]. Copyright 2016, Springer Nature. (**i**) Reconfigurable chiral plasmonic metamolecules fabricated using pH-sensitive dynamic DNA origami [46]. Copyright 2017, AAAS. (**j**) Dynamic modulation of AuNP lattice configurations through reconfigurable pH-responsive DNA origami lattices [47]. Copyright 2023, American Chemical Society.

https://doi.org/10.3390/molecules31030513

4. Tunable Plasmonic Properties of DNA Nanostructure-Mediated Metal Nanoparticle Assemblies

The integration of metallic nanoparticles with DNA nanostructures has established a powerful paradigm for plasmonic material design, in which the structural programmability of DNA scaffolds is directly translated into tunable optical responses. By enabling nanometer-scale control over interparticle distance, spatial orientation, symmetry, and local dielectric environment, DNA nanostructures provide a unique platform for engineering plasmonic coupling phenomena that are difficult to access using conventional top-down fabrication strategies. This precise control over nanoscale geometry allows collective plasmonic modes to be deliberately designed and optimized, giving rise to well-defined optical outputs.

As a result, key plasmonic signal modalities relevant to biosensing, including CD, fluorescence modulation, and SERS can be systematically regulated through rational structural design. Importantly, DNA nanostructures can function either as static architectural frameworks that maximize electromagnetic field enhancement or, when combined with reconfigurable motifs, as dynamic platforms capable of converting molecular inputs into optical signals. Together, these DNA nanostructure-based plasmonic systems enable biosensing architectures that combine strong signal amplification with programmable optical signal transduction [48–50].

4.1. Chiroptical Modulation

Chiroptical responses are among the most geometry-sensitive plasmonic phenomena, making them particularly suitable for elucidating how DNA nanostructures mediate plasmonic mode engineering. DNA nanostructures enable the precise construction of chiral plasmonic assemblies by offering nanometer-level control over nanoparticle positioning, relative orientation, and global symmetry. This structural precision allows the realization of a broad range of geometrically chiral architectures, including helices, asymmetrically positioned dimers, and cross-stacked oligomers, which collectively give rise to pronounced plasmonic CD signals in the visible and near-infrared regions.

A representative example is provided by the work of Urban and co-workers, who arranged gold nanoparticles into well-defined left- or right-handed helices on ring-shaped DNA origami templates to construct plasmonic toroidal metamolecules. Finite-element simulations faithfully reproduced the experimentally observed CD spectra and provided mechanistic insight into their physical origin. Analysis of the induced charge and current distributions revealed that, at plasmonic resonance, strong electromagnetic coupling between adjacent nanoparticles leads to the formation of collective plasmonic modes. The symmetry and handedness of these modes are dictated by the DNA-encoded chiral geometry, resulting in distinct interactions with left- and right-circularly polarized light and, consequently, pronounced CD responses (Figure 3a) [51]. These studies establish a clear structure–property relationship linking DNA nanostructures programmed geometry, collective plasmonic coupling, and emergent chiroptical activity.

Beyond purely geometric effects, DNA nanostructure-templated chiral assemblies also offer opportunities for further tuning chiroptical responses through material and hybridization strategies. For example, coating gold nanoparticles with silver shells enables precise adjustment of plasmonic resonance energies, while incorporation of excitonic components such as J-aggregate dyes introduce plasmon–exciton coupling (Figure 3b) [52,53]. In such hybrid systems, the interplay between chiral plasmonic modes and excitonic transitions leads to enhanced and spectrally tunable CD signals, highlighting how DNA nanostructures provide a modular framework for engineering complex chiroptical responses.

https://doi.org/10.3390/molecules31

While many of these studies focus on static chiral architectures to maximize CD signal strength, the same design principles can be extended to reconfigurable DNA scaffolds, in which controlled structural rearrangements translate into switchable chiroptical outputs [54,55]. This capability forms the basis for using plasmonic CD as an optical readout in responsive sensing systems.

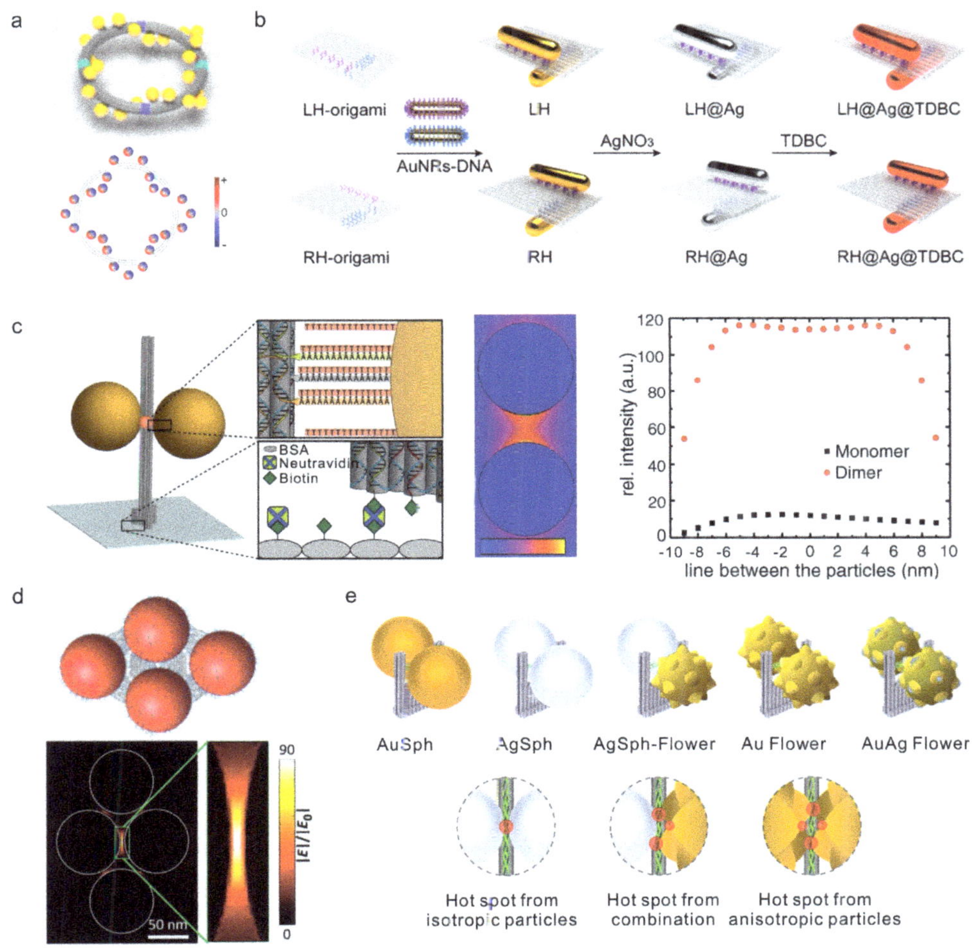

Figure 3. Tunable optical properties of DNA plasmonic nanostructure assemblies. (**a**) DNA origami-templated gold nanoparticle helices exhibiting strong CD due to geometry-controlled collective plasmonic coupling [51]. Copyright 2016, American Chemical Society. (**b**) Chiral plasmonic–excitonic hybrids, where gold nanorod dimers on DNA scaffolds interact with excitonic dyes to produce enhanced, tunable CD responses. The purple and blue lines represent the assembly sites of gold nanorod [53]. Copyright 2021, American Chemical Society. (**c**) DNA origami nanoantennas positioning fluorophores in nanoparticle dimers, generating strong and reproducible fluorescence enhancement at electromagnetic hotspots [56]. Copyright 2012, AAAS. (**d**) Tetrameric gold nanoparticle assemblies on DNA origami forming Fano-resonant SERS hotspots, enabling quantitative Raman measurements with controlled reporter placement [41]. Copyright 2019, AAAS. (**e**) SERS optimization through DNA-directed control of nanoparticle composition and morphology, showing how geometric precision and material choice enhance local field confinement [57]. Copyright 2023, American Chemical Society.

https://doi.org/10.3390/molecules31030513

4.2. Fluorescence Enhancement and Energy Transfer Control

Fluorescence-based plasmonic readouts offer a complementary perspective by probing highly localized electromagnetic field enhancement at the single-emitter level. DNA nanostructures serve as precise scaffolds for positioning fluorophores within plasmonic nanocavities, enabling systematic regulation of excitation rates, emission intensity, polarization, and directionality through controlled light-matter interactions.

DNA origami-based plasmonic nanoantennas, typically composed of gold nanoparticle or nanorod dimers, have been extensively employed to investigate fluorescence enhancement mechanisms (Figure 3c). By placing fluorophores at predefined docking sites within nanogaps, local electromagnetic fields can be strongly enhanced, leading to fluorescence enhancement factors exceeding two orders of magnitude [56]. Such static architectures function as efficient optical amplifiers, providing reproducible and quantifiable enhancement of weak fluorescence signals.

Beyond excitation enhancement, plasmonic nanocavities also modify emission dynamics by reshaping the local density of optical states [58]. Single-molecule measurements reveal pronounced reductions in fluorescence lifetime, indicative of Purcell-enhanced radiative decay. Importantly, the balance between radiative enhancement and nonradiative energy dissipation is critically dependent on emitter–metal separation. The nanometer-scale distance control afforded by DNA origami enables fluorophores to be positioned within an optimal regime where radiative processes dominate over quenching, allowing predictable and rational design of fluorescence-modulating plasmonic structures.

Although most fluorescence enhancement studies employ static DNA scaffolds to maximize signal amplification, incorporation of stimulus-responsive DNA elements enables reversible modulation of emitter–nanoparticle coupling. In such cases, fluorescence intensity and lifetime can serve as optical reporters of DNA conformational states, providing a direct link between nanoscale structural changes and measurable optical outputs.

4.3. Surface-Enhanced Raman Scattering

Surface-enhanced Raman scattering represents an extreme manifestation of plasmonic field confinement, relying on highly localized electromagnetic hotspots to amplify intrinsically weak molecular vibrational signals. DNA origami provides an exceptional platform for SERS studies by enabling precise control over hotspot geometry, plasmonic mode structure, and molecular positioning [49,59,60].

As demonstrated by Fang and co-workers, gold nanoparticles can be programmably assembled into well-defined tetrameric superstructures using DNA origami super-frame designs (Figure 3d) [41]. These assemblies support Fano resonances arising from interference between super-radiant bright modes and sub-radiant dark modes. Electromagnetic simulations based on finite-difference time-domain and finite-element methods reveal that, at the Fano dip wavelength, radiative losses are suppressed and electromagnetic energy is strongly confined within interparticle gaps, resulting in extremely high local field enhancements. Crucially, the addressability of DNA origami allows a defined number of Raman reporter molecules to be positioned precisely within these computationally identified hotspots. Experimentally, SERS intensities exhibit stepwise, quantized increases with increasing molecular occupancy, directly linking signal strength to molecular number. This level of control transforms SERS from a stochastic enhancement phenomenon into a predictable and quantitative process governed by DNA-programmed plasmonic architecture.

Additional studies further demonstrate that nanoparticle composition and morphology play critical roles in determining SERS performance. Systematic comparisons of gold, silver, and anisotropic nanostructures assembled on DNA origami, supported by large-scale statistical analyses and electromagnetic simulations, show that plasmonic resonance

https://doi.org/10.3390/molecules310

positions and local field enhancements can be optimized through rational material selection and precise geometric control (Figure 3e) [57]. In this context, DNA nanostructures act as nanoscale photonic engineering platforms that enable reproducible and high-performance SERS substrates.

Taken together, these studies demonstrate that DNA nanostructures enable plasmonic properties to be engineered at a mechanistic level through precise control over nanoscale geometry and electromagnetic coupling. By enforcing well-defined interparticle spacing, orientation, and symmetry, DNA-guided assemblies improve the reproducibility of plasmonic architectures and, consequently, the consistency of their optical signal outputs. While different optical modalities exhibit varying sensitivity to nanoscale precision, a clear correlation between structural fidelity and signal reproducibility is generally observed. Collective chiroptical signals are relatively tolerant to minor geometric variations, whereas SERS readouts impose stricter requirements on hotspot definition.

Within this framework, static DNA-templated plasmonic architectures primarily function as efficient signal amplifiers by maximizing electromagnetic field enhancement, while the incorporation of dynamic DNA nanotechnology introduces responsiveness, allowing plasmonic systems to actively transduce molecular inputs into optical outputs. The integration of static amplification and dynamic signal transduction provides a versatile foundation for biosensing, enabling molecular recognition events to be both strongly amplified and reliably converted into quantitative optical readouts.

5. Biosensing Applications of DNA Nanostructure-Mediated Metal Nanoparticle Assemblies

The integration of DNA nanostructures with metal nanoparticles provides a versatile strategy for translating molecular recognition events into highly sensitive optical signals. Owing to their inherent programmability, DNA nanostructures enable the precise spatial organization of metal nanoparticles and optical emitters, yielding assemblies with finely tunable chiroptical, fluorescence, and Raman responses. Importantly, these optical signals can be dynamically modulated in response to specific biochemical interactions, allowing DNA nanostructure-based plasmonic systems to function as powerful transducers for molecular diagnostics and environmental biosensing [61,62].

5.1. Chirality-Based Biosensing

DNA nanostructure-templated chiral plasmonic assemblies are particularly effective at converting molecular recognition events into measurable variations in CD, providing a label-free and real-time sensing modality. By arranging AuNPs or AuNRs on DNA origami scaffolds with well-defined asymmetric, twisted, or crossed geometries, controllable chiroptical signals can be generated. Hybridization with complementary nucleic acid targets or binding of aptamer–ligand pairs induce conformational changes in the DNA framework, thereby modulating plasmonic coupling and producing distinct CD intensity or sign variations.

Using this principle, twisted AuNR dimers assembled through DNA hybridization have been shown to generate CD responses that correlate linearly with target DNA concentration, achieving detection limits as low as 3.7 aM [63]. Extending this concept to intracellular environments, side-by-side AuNR assemblies positioned within living cells enabled real-time monitoring of microRNA recognition via in situ CD signal changes, demonstrating the feasibility of chiroptical sensing in complex biological contexts [64].

Beyond static sensing, reconfigurable chiral plasmonic systems based on DNA origami have demonstrated broad applicability for detecting diverse targets, including nucleic acids, proteins, and small molecules [65–67]. Upon specific molecular recognition, these archi-

https://doi.org/10.3390/molecules31030513

tectures undergo programmable conformational transitions that alter their chiral configuration, yielding quantifiable CD readouts. For example, cross-arranged AuNR dimers on DNA origami scaffolds exhibited reversible chirality switching upon viral RNA hybridization, enabling quantitative detection with limits of approximately 100 pM [65]. Similarly, aptamer-integrated chiral origami assemblies have been engineered to recognize ligands such as adenosine and thrombin, where target binding induces structural reorganization of the plasmonic framework, with dissociation constants down to 1.4 nM [67].

To further enhance sensitivity toward low-abundance biomarkers, signal amplification modules have been integrated upstream of the chiroptical readout. In representative designs, DNA amplification circuits recycle trace amounts of target molecules to generate abundant trigger strands, which subsequently induce large-scale reconfiguration of AuNR-functionalized DNA origami nanotweezers. This strategy effectively amplifies weak molecular recognition events into macroscopic CD responses, enabling detection of tumor-related mRNAs, adenosine, and cell-surface proteins at picomolar concentrations or even at the single-cell level (Figure 4a) [68]. Together, these examples highlight the potential of DNA nanostructure-based chiral plasmonic systems for ultrasensitive, real-time bioanalysis.

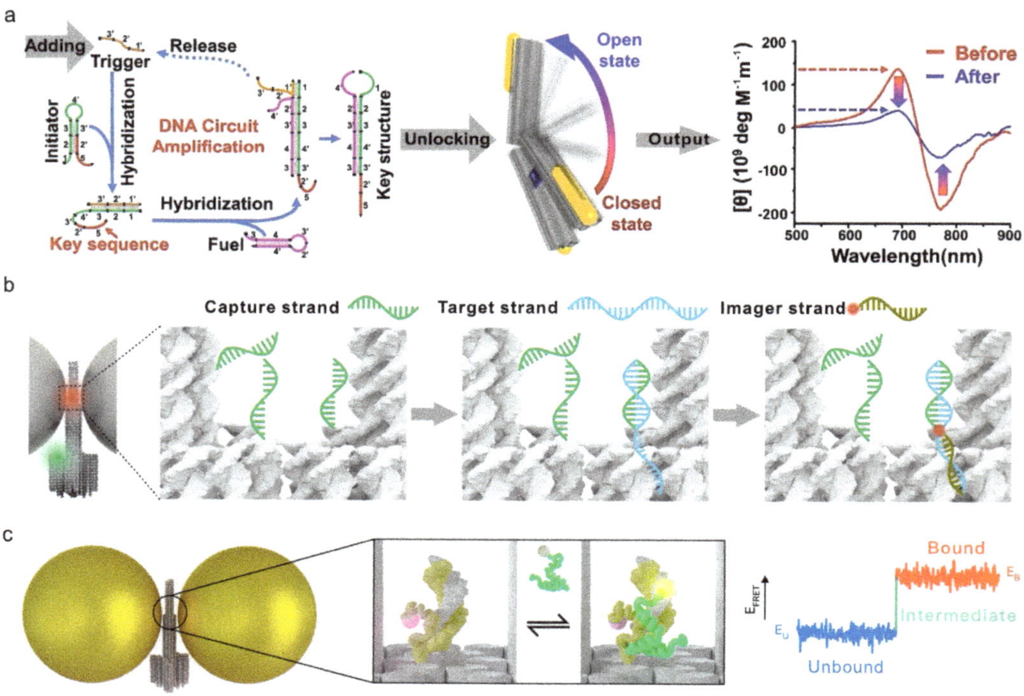

Figure 4. Chirality- and fluorescence-based biosensing. (**a**) A DNA amplification circuit converts the target analyte into a key structure, triggering a conformational transformation of the plasmonic assembly and generating a corresponding CD signal. The red line segment represents the key sequence that interacts with the nanotweezers [68]. Copyright 2022, Wiley. (**b**) Capture strands immobilized on a nanoantenna recognize and bind target molecules, exposing a hybridization site for an imaging strand that produces a radiometric fluorescence signal [69]. Copyright 2021, Springer Nature. (**c**) DNA origami-based nanoantennas enable single-molecule investigations of the coupled folding and binding dynamics of proteins [70]. Copyright 2024, American Chemical Society.

https://doi.org/10.3390/molecules31

5.2. Fluorescence-Based Biosensing

The nanometer-scale spatial control afforded by DNA nanostructures provides an ideal platform for plasmon-enhanced fluorescence biosensing, where the radiative properties of fluorophores can be precisely modulated. DNA templates enable the construction of optical nanoantennas that translate molecular recognition events into amplified fluorescence signals.

Seminal work by Tinnefeld and co-workers established DNA origami-based optical nanoantennas as robust systems for plasmon-enhanced fluorescence sensing. By systematically optimizing nanoparticle size, shape, material composition, and fluorophore positioning, these platforms enabled tunable fluorescence enhancement and single-molecule nucleic acid detection [50,69,71]. In a representative design, single 80 nm Ag nanoparticles were anchored on DNA origami templates in a molecular beacon-like configuration, resulting in fluorescence enhancement of several orders of magnitude upon target hybridization. Using this approach, Zika virus DNA and RNA sequences were sensitively detected directly in serum samples [71].

Further refinement led to the development of trident-shaped NanoAntennas with Cleared Hotspots (NACHOS), in which two Ag nanopillars defined a nanoscale junction containing capture strands. Upon target hybridization, the hotspot became accessible to an imager strand, yielding fluorescence enhancement factors of approximately 467-fold. Remarkably, the resulting signal could be visualized using a smartphone-based imaging setup, enabling portable single-molecule detection [69]. This strategy was subsequently applied to quantify antibiotic-resistance markers from *Klebsiella pneumoniae* at attomolar concentrations directly in human plasma, achieving polymerase chain reaction-level sensitivity without enzymatic amplification (Figure 4b). Beyond nucleic acid sensing, DNA–metal nanoantennas have been adapted for small-molecule detection by incorporating aptamer recognition elements. For example, ATP-binding aptamers integrated into DNA origami-templated Au@Ag nanostructures undergo target-induced conformational switching to form G-quadruplex structures, producing strong fluorescence enhancement and enabling single-molecule ATP detection [72].

In addition to static sensing, plasmonic DNA nanoantennas have enabled real-time observation of biomolecular dynamics. By confining fluorophores within plasmonic hotspots, photon emission rates are significantly increased, allowing acquisition of single-molecule FRET trajectories with temporal resolution improved by approximately one order of magnitude. This capability has facilitated direct observation of rapid folding transitions in intrinsically disordered proteins and short DNA duplexes, providing experimental access to transient states previously accessible mainly through simulations (Figure 4c) [70]. Collectively, these studies underscore the versatility of DNA nanostructure-mediated fluorescence platforms for both ultrasensitive biosensing and dynamic molecular analysis.

5.3. SERS-Based Biosensing

SERS-based biosensing benefits substantially from DNA-guided assembly strategies, which enable rational design of plasmonic nanostructures with programmable geometries and well-defined nanogaps. The structural precision of DNA nanotechnology allows fine control over gap size, nanoparticle composition, and analyte positioning, resulting in ultrasensitive, reproducible, and multiplexed Raman-based detection.

For nucleic acid analysis, DNA nanostructure-assembled SERS platforms have incorporated enzymatic target-recycling strategies to overcome sensitivity limitations. Duplex-specific nuclease-assisted SERS probes and exosome-targeted DNA–Au@Ag conjugates have been applied for the detection of miRNA-10b in serum and exosomal samples, achieving attomolar to femtomolar detection limits with high sequence specificity [73]. DNA

https://doi.org/10.3390/molecules31030513

nanostructure engineering has further enabled multiplexed miRNA profiling, where tetrahedral DNA probes or asymmetric core–shell nanoparticles functionalized with distinct Raman reporters allow simultaneous detection of multiple miRNA targets within a single sample [74].

Beyond nucleic acids, DNA origami provides a programmable scaffold for assembling plasmonic nanoantennas responsive to small molecules and proteins. Aptamer-functionalized origami templates decorated with AuNPs or AgNPs can transduce ligand binding into structural rearrangement, producing pronounced Raman signal modulation over wide concentration ranges (10^{-10}–10^{-5} M) with sub-nanomolar detection limits (Figure 5a) [75,76]. Moreover, DNA-templated assembly of sharp-featured nanostructures, such as bimetallic Au@Ag nanostars, exploits strong tip-enhanced fields to enable sensitive detection of metabolites like pyocyanin at clinically relevant concentrations [59].

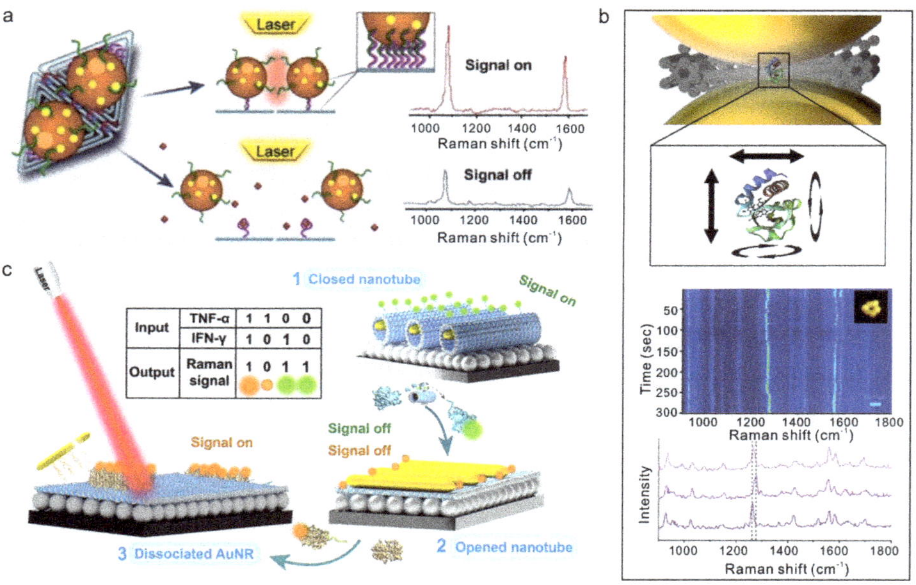

Figure 5. SERS-based biosensing (**a**) A DNA origami-assembled nanoantenna-based SERS biosensor for the ultrasensitive detection of trace amounts of diethylstilbestrol [75]. Copyright 2025, Royal Society of Chemistry. (**b**) A single-molecule SERS platform based on DNA origami nanoantennas for probing the dynamic conformational transitions of cytochrome c. [77]. Copyright 2024, American Chemical Society. (**c**) A programmable DNA origami plasmonic nanoantenna for multiplex SERS detection of TNF-α and IFN-γ relevant to cancer immunotherapy [78]. Copyright 2024, American Chemical Society.

DNA origami-defined plasmonic hotspots have also enabled label-free single-protein identification and enzymatic analysis. By precisely positioning proteins such as thrombin or epidermal growth factor receptor within nanogaps, distinct single-molecule Raman fingerprints can be acquired [79]. Furthermore, real-time monitoring of enzymatic reactions has been achieved by immobilizing cytochrome C or horseradish peroxidase within DNA-defined hotspots, allowing catalytic processes to be tracked through Raman spectral evolution (Figure 5b) [77,80].

The programmability of DNA origami additionally facilitates multiplexed and logic-based biosensing. Stimuli-responsive nanoantennas incorporating Au nanorods and aptamer-functionalized origami sheets have been designed to detect cytokines such as

https://doi.org/10.3390/molecules310

TNF-α and IFN-γ, with Raman signal variations serving as Boolean logic outputs for immune-response profiling (Figure 5c) [78]. Integration of multiple optical modalities further enhances analytical robustness. For example, Y-DNA-mediated assembly of chiral AuNP core–satellite structures generate correlated SERS and CD signals, enabling dual-mode miRNA quantification in living cells [81]. Such multimodal platforms illustrate how DNA-assembled plasmonic nanostructures combine structural programmability, optical enhancement, and biochemical specificity to advance next-generation molecular diagnostics.

6. Conclusions and Future Perspectives

DNA nanostructure-guided assemblies of metal nanoparticles have emerged as a versatile and powerful platform for biosensing applications. The intrinsic programmability of DNA nanostructures allows precise spatial organization of nanoparticles, while dynamic scaffolds enable stimulus-responsive reconfiguration. Together, these features provide highly controlled, multifunctional, and tunable platforms capable of enhancing signal transduction, enabling multimodal detection, and implementing logic- or condition-dependent sensing.

Despite these remarkable advances, several challenges remain. The stability of DNA-nanoparticle assemblies under complex biological conditions, including nucleases, high ionic strength, or extreme pH, still requires further optimization. Large-scale, reproducible fabrication of uniform assemblies, particularly for hierarchical or multi-component architectures, remains technically demanding. In addition, integrating multiple signal transduction modalities within a single platform, while maintaining predictable and non-interfering outputs, presents both design and experimental challenges.

Looking forward, several opportunities are evident. Advancements in dynamic and responsive DNA nanostructures may enable more sophisticated control over nanoparticle arrangement, facilitating smart biosensing systems capable of autonomous signal processing and decision-making. Integration with microfluidic, wearable, or implantable devices could expand the practical applications of DNA nanostructure-guided nanoparticle biosensors in point-of-care diagnostics, environmental monitoring, and real-time cellular sensing. Moreover, combining DNA nanostructure-guided nanoparticle assemblies with other nanomaterials, such as quantum dots, upconversion nanoparticles, or 2D materials, may further enhance multimodal detection capabilities and functional versatility.

In summary, DNA nanostructure-assembled metal nanoparticles provide a unique combination of structural programmability, dynamic reconfigurability, and multifunctional signal transduction, offering a robust foundation for the next generation of biosensing technologies. Continued advances in scaffold design, assembly strategies, and integration with diverse signal modalities are expected to drive the development of highly sensitive, selective, and programmable biosensors for biomedical and environmental applications.

Funding: This work was supported by the National Key R&D Program of China (2023YFB3208200), the National Natural Science Foundation of China (22393931).

Institutional Review Board Statement: Not applicable.

Informed Consent Statement: Not applicable.

Data Availability Statement: No new data were created or analyzed in this study. Data sharing is not applicable.

Conflicts of Interest: The authors declare no conflicts of interest.

https://doi.org/10.3390/molecules31030513

References

1. Wu, Y.; Xu, X.; Zhu, Y.C.; Wan, J.X.; Wang, X.B.; Zhou, X.; Li, X.J.; Zhou, W.D. Research Progress on Multiplexed Path Detection Using Optical Biosensors. *Biosensors* **2025**, *15*, 378. [CrossRef]
2. Joshi, K.M.; Salve, S.; Dhanwade, D.; Chavhan, M.; Jagtap, S.; Shinde, M.; Holkar, R.; Patil, R.; Chabukswar, V. Advancing p biosensors: Redefining detection through innovations in materials, mechanisms, and applications for precision medicin global diagnostics. *RSC Adv.* **2025**, *15*, 11523–11536. [CrossRef]
3. Huynh, K.H.; Pham, X.H.; Kim, J.; Lee, S.H.; Chang, H.; Rho, W.Y.; Jun, B.H. Synthesis, Properties, and Biological Applicati Metallic Alloy Nanoparticles. *Int. J. Mol. Sci.* **2020**, *21*, 5174. [CrossRef]
4. Báez, D.F.; Gallardo-Toledo, E.; Oyarzún, M.P.; Araya, E.; Kogan, M.J. The Influence of Size and Chemical Composition of and Gold Nanoparticles on in vivo Toxicity with Potential Applications to Central Nervous System Diseases. *Int. J. Nan* **2021**, *16*, 2187–2201. [CrossRef]
5. Tim, B.; Blaszkiewicz, P.; Kotkowiak, M. Recent Advances in Metallic Nanoparticle Assemblies for Surface-Enhanced Spectro *Int. J. Mol. Sci.* **2022**, *23*, 291. [CrossRef]
6. Chen, X.R.; Ding, L.; Huang, X.L.; Xiong, Y.H. Tailoring noble metal nanoparticle designs to enable sensitive latera immunoassay. *Theranostics* **2022**, *12*, 574–602. [CrossRef]
7. Balena, A.; D'Amato, M.; Kashif, M.F.; Ding, C.J.; Belzane, L.; De Marzo, G.; Le Jeannic, H.; De Vittorio, M.; Pisanello, F.; Br A. Deterministic Fabrication of Plasmonic Nanostructures on Optical Nanofibers via Blurred Electron Beam Deposition. *Ad* **2025**, *12*, e07004. [CrossRef]
8. Jawad, M.S.; Rahman, W.U.; Khan, M.; Faysal, A.M. Nanomaterial assembly pathways: Comparative insights into self-ass and directed assembly techniques. *Next Mater.* **2026**, *10*, 101422. [CrossRef]
9. Barad, H.N.; Kwon, H.; Alarcón-Correa, M.; Fischer, P. Large Area Patterning of Nanoparticles and Nanostructures: Cu Status and Future Prospects. *ACS Nano* **2021**, *15*, 5861–5875. [CrossRef]
10. Jambhulkar, S.; Ravichandran, D.; Zhu, Y.X.; Thippanna, V.; Ramanathan, A.; Patil, D.; Fonseca, N.; Thummalapalli Sundaravadivelan, B.; Sun, A.L.; et al. Nanoparticle Assembly: From Self-Organization to Controlled Micropatterni Enhanced Functionalities. *Small* **2024**, *20*, e2306394. [CrossRef]
11. Seeman, N.C. Nucleic acid junctions and lattices. *J. Theor. Biol.* **1982**, *99*, 237–247. [CrossRef] [PubMed]
12. Kallenbach, N.R.; Ma, R.-I.; Seeman, N.C. An immobile nucleic acid junction constructed from oligonucleotides. *Nature* **198** 829–831. [CrossRef]
13. Mathieu, F.; Liao, S.; Kopatsch, J.; Wang, T.; Mao, C.; Seeman, N.C. Six-Helix Bundles Designed from DNA. *Nano Lett.* **2** 661–665. [CrossRef] [PubMed]
14. He, Y.; Ye, T.; Su, M.; Zhang, C.; Ribbe, A.E.; Jiang, W.; Mao, C. Hierarchical self-assembly of DNA into symmetric supramol polyhedra. *Nature* **2008**, *452*, 198–201. [CrossRef]
15. He, Y.; Chen, Y.; Liu, H.; Ribbe, A.E.; Mao, C. Self-Assembly of Hexagonal DNA Two-Dimensional (2D) Arrays. *J. Am. Che* **2005**, *127*, 12202–12203. [CrossRef]
16. He, Y.; Tian, Y.; Chen, Y.; Deng, Z.; Ribbe, A.E.; Mao, C. Sequence symmetry as a tool for designing DNA nanostructures. *Chem. Int. Ed.* **2005**, *44*, 6694–6696. [CrossRef]
17. He, Y.; Tian, Y.; Ribbe, A.E.; Mao, C. Highly Connected Two-Dimensional Crystals of DNA Six-Point-Stars. *J. Am. Chem. Soc* *128*, 15978–15979. [CrossRef]
18. Rothemund, P.W.K. Folding DNA to create nanoscale shapes and patterns. *Nature* **2006**, *440*, 297–302. [CrossRef]
19. Andersen, E.S.; Dong, M.; Nielsen, M.M.; Jahn, K.; Subramani, R.; Mamdouh, W.; Golas, M.M.; Sander, B.; Stark, H.; Ol C.L.P.; et al. Self-assembly of a nanoscale DNA box with a controllable lid. *Nature* **2009**, *459*, 73–76. [CrossRef]
20. Dietz, H.; Douglas, S.M.; Shih, W.M. Folding DNA into Twisted and Curved Nanoscale Shapes. *Science* **2009**, *325*, 72 [CrossRef]
21. Douglas, S.M.; Dietz, H.; Liedl, T.; Hoegberg, B.; Graf, F.; Shih, W.M. Self-assembly of DNA into nanoscale three-dimen shapes. *Nature* **2009**, *459*, 414–418. [CrossRef]
22. Han, D.; Pal, S.; Nangreave, J.; Deng, Z.; Liu, Y.; Yan, H. DNA Origami with Complex Curvatures in Three-Dimensional *Science* **2011**, *332*, 342–346. [CrossRef]
23. Iinuma, R.; Ke, Y.; Jungmann, R.; Schlichthaerle, T.; Woehrstein, J.B.; Yin, P. Polyhedra Self-Assembled from DNA Tripoc Characterized with 3D DNA-PAINT. *Science* **2014**, *344*, 65–69. [CrossRef] [PubMed]
24. Zhang, F.; Jiang, S.; Wu, S.; Li, Y.; Mao, C.; Liu, Y.; Yan, H. Complex wireframe DNA origami nanostructures with mul junction vertices. *Nat. Nanotechnol.* **2015**, *10*, 779–784. [CrossRef] [PubMed]
25. Zhao, Y.; Shi, L.X.; Kuang, H.; Xu, C.L. DNA-Driven Nanoparticle Assemblies for Biosensing and Bioimaging. *Top. Curr.* **2020**, *378*, 18. [CrossRef] [PubMed]
26. He, Q.Y.; Wu, Q.; Feng, X.R.; Liao, Z.Y.; Peng, W.Y.; Liu, Y.F.; Peng, D.M.; Liu, Z.B.; Mo, M. Interfacing DNA with nanopa Surface science and its applications in biosensing. *Int. J. Biol. Macromol.* **2020**, *151*, 757–780. [CrossRef]

https://doi.org/10.3390/molecules31

Chen, C.; Wei, X.F.; Parsons, M.F.; Guo, J.J.; Banal, J.L.; Zhao, Y.N.; Scott, M.N.; Schlau-Cohen, G.S.; Hernandez, R.; Bathe, M. Nanoscale 3D spatial addressing and valence control of quantum dots using wireframe DNA origami. *Nat. Commun.* **2022**, *13*, 4935. [CrossRef]
Dey, S.; Fan, C.; Gothelf, K.V.; Li, J.; Lin, C.; Liu, L.; Liu, N.; Nijenhuis, M.A.D.; Saccà, B.; Simmel, F.C.; et al. DNA origami. *Nat. Rev. Methods Primers* **2021**, *1*, 13. [CrossRef]
Qian, L.; Wang, Y.; Zhang, Z.; Zhao, J.; Pan, D.; Zhang, Y.; Liu, Q.; Fan, C.; Hu, J.; He, L. Analogic China map constructed by DNA. *Chin. Sci. Bull.* **2006**, *51*, 2973–2976. [CrossRef]
Han, D.; Pal, S.; Liu, Y.; Yan, H. Folding and cutting DNA into reconfigurable topological nanostructures. *Nat. Nanotechnol.* **2010**, *5*, 712–717. [CrossRef]
Li, G.Z.; Chen, C.Y.; Li, Y.Y.; Wang, B.; Wen, J.L.; Guo, M.Y.; Chen, M.; Zhang, X.B.; Ke, G.L. DNA-Origami-Based Precise Molecule Assembly and Their Biological Applications. *Nano Lett.* **2024**, *24*, 11335–11348. [CrossRef]
Kearney, C.J.; Lucas, C.R.; O'Brien, F.J.; Castro, C.E. DNA Origami: Folded DNA-Nanodevices That Can Direct and Interpret Cell Behavior. *Adv. Mater.* **2016**, *28*, 5509–5524. [CrossRef] [PubMed]
Kuzyk, A.; Yang, Y.; Duan, X.; Stoll, S.; Govorov, A.O.; Sugiyama, H.; Endo, M.; Liu, N. A light-driven three-dimensional plasmonic nanosystem that translates molecular motion into reversible chiroptical function. *Nat. Commun.* **2016**, *7*, 10591. [CrossRef] [PubMed]
Ijäs, H.; Hakaste, I.; Shen, B.; Kostiainen, M.A.; Linko, V. Reconfigurable DNA Origami Nanocapsule for pH-Controlled Encapsulation and Display of Cargo. *ACS Nano* **2019**, *13*, 5959–5967. [CrossRef] [PubMed]
Suzuki, Y.; Kawamata, I.; Mizuno, K.; Murata, S. Large Deformation of a DNA-Origami Nanoarm Induced by the Cumulative Actuation of Tension-Adjustable Modules. *Angew. Chem. Int. Ed.* **2020**, *59*, 6230–6234. [CrossRef]
Wang, D.; Yu, L.; Huang, C.-M.; Arya, G.; Chang, S.; Ke, Y. Programmable Transformations of DNA Origami Made of Small Modular Dynamic Units. *J. Am. Chem. Soc.* **2021**, *143*, 2256–2263. [CrossRef]
Kühler, P.; Roller, E.-M.; Schreiber, R.; Liedl, T.; Lohmüller, T.; Feldmann, J. Plasmonic DNA-Origami Nanoantennas for Surface-Enhanced Raman Spectroscopy. *Nano Lett.* **2014**, *14*, 2914–2919. [CrossRef]
Pal, S.; Deng, Z.; Ding, B.; Yan, H.; Liu, Y. DNA-Origami-Directed Self-Assembly of Discrete Silver-Nanoparticle Architectures. *Angew. Chem. Int. Ed.* **2010**, *49*, 2700–2704. [CrossRef]
Pal, S.; Deng, Z.; Wang, H.; Zou, S.; Liu, Y.; Yan, H. DNA Directed Self-Assembly of Anisotropic Plasmonic Nanostructures. *J. Am. Chem. Soc.* **2011**, *133*, 17606–17609. [CrossRef]
Gür, F.N.; McPolin, C.P.T.; Raza, S.; Mayer, M.; Roth, D.J.; Steiner, A.M.; Löffler, M.; Fery, A.; Brongersma, M.L.; Zayats, A.V.; et al. DNA-Assembled Plasmonic Waveguides for Nanoscale Light Propagation to a Fluorescent Nanodiamond. *Nano Lett.* **2018**, *18*, 7323–7329. [CrossRef]
Fang, W.; Jia, S.; Chao, J.; Wang, L.; Duan, X.; Liu, H.; Li, Q.; Zuo, X.; Wang, L.; Wang, L.; et al. Quantizing single-molecule surface-enhanced Raman scattering with DNA origami metamolecules. *Sci. Adv.* **2019**, *5*, eaau4506. [CrossRef]
Dong, B.Q.; Xu, X.; Guan, R.C.; Jiang, S.X.; Ma, L.; Hu, H.T.; Ke, Y.G.; Liu, N.; Lan, X. Two-Dimensional, Chiral Colloidal Superlattices Engineered with DNA Origami. *Nano Lett.* **2025**, *25*, 5705–5712. [CrossRef]
Kuzyk, A.; Schreiber, R.; Fan, Z.; Pardatscher, G.; Roller, E.-M.; Högele, A.; Simmel, F.C.; Govorov, A.O.; Liedl, T. DNA-based self-assembly of chiral plasmonic nanostructures with tailored optical response. *Nature* **2012**, *483*, 311–314. [CrossRef] [PubMed]
Liu, N.; Liedl, T. DNA-Assembled Advanced Plasmonic Architectures. *Chem. Rev.* **2018**, *118*, 3032–3053. [CrossRef] [PubMed]
Zhou, C.; Duan, X.; Liu, N. A plasmonic nanorod that walks on DNA origami. *Nat. Commun.* **2015**, *6*, 8102. [CrossRef] [PubMed]
Kuzyk, A.; Urban, M.J.; Idili, A.; Ricci, F.; Liu, N. Selective control of reconfigurable chiral plasmonic metamolecules. *Sci. Adv.* **2017**, *3*, e1602803. [CrossRef]
Julin, S.; Linko, V.; Kostiainen, M.A. Reconfigurable pH-Responsive DNA Origami Lattices. *ACS Nano* **2023**, *17*, 11014–11022. [CrossRef]
Schlichthaerle, T.; Strauss, M.T.; Schueder, F.; Woehrstein, J.B.; Jungmann, R. DNA nanotechnology and fluorescence applications. *Curr. Opin. Biotechnol.* **2016**, *39*, 41–47. [CrossRef]
Mehmandoust, S.; Eskandari, V.; Karooby, E. A Review of Fabrication of DNA Origami Plasmonic Structures for the Development of Surface-Enhanced Raman Scattering (SERS) Platforms. *Plasmonics* **2023**, *19*, 1131–1143. [CrossRef]
Glembockyte, V.; Grabenhorst, L.; Trofymchuk, K.; Tinnefeld, P. DNA Origami Nanoantennas for Fluorescence Enhancement. *Acc. Chem. Res.* **2021**, *54*, 3338–3348. [CrossRef]
Urban, M.J.; Dutta, P.K.; Wang, P.F.; Duan, X.Y.; Shen, X.B.; Ding, B.Q.; Ke, Y.G.; Liu, N. Plasmonic Toroidal Metamolecules Assembled by DNA Origami. *J. Am. Chem. Soc.* **2016**, *138*, 5495–5498. [CrossRef]
Kneer, L.M.; Roller, E.M.; Besteiro, L.V.; Schreiber, R.; Govorov, A.O.; Liedl, T. Circular Dichroism of Chiral Molecules in DNA-Assembled Plasmonic Hotspots. *ACS Nano* **2018**, *12*, 9110–9115. [CrossRef] [PubMed]
Zhu, J.; Wu, F.; Han, Z.; Shang, Y.; Liu, F.; Yu, H.; Yu, L.; Li, N.; Ding, B. Strong Light-Matter Interactions in Chiral Plasmonic-Excitonic Systems Assembled on DNA Origami. *Nano Lett.* **2021**, *21*, 3573–3580. [CrossRef] [PubMed]

54. Zhou, C.; Duan, X.; Liu, N. DNA-Nanotechnology-Enabled Chiral Plasmonics: From Static to Dynamic. *Acc. Chem. Res.* **201** 2906–2914. [CrossRef] [PubMed]
55. Wang, M.; Dong, J.; Zhou, C.; Xie, H.; Ni, W.; Wang, S.; Jin, H.; Wang, Q. Reconfigurable Plasmonic Diastereomers Assemble DNA Origami. *ACS Nano* **2019**, *13*, 13702–13708. [CrossRef]
56. Acuna, G.P.; Moller, F.M.; Holzmeister, P.; Beater, S.; Lalkens, B.; Tinnefeld, P. Fluorescence enhancement at docking sit DNA-directed self-assembled nanoantennas. *Science* **2012**, *338*, 506–510. [CrossRef]
57. Kanehira, Y.; Tapio, K.; Wegner, G.; Kogikoski, S., Jr.; Rustig, S.; Prietzel, C.; Busch, K.; Bald, I. The Effect of Nanopa Composition on the Surface-Enhanced Raman Scattering Performance of Plasmonic DNA Origami Nanoantennas. *ACS N* **2023**, *17*, 21227–21239. [CrossRef]
58. Chikkaraddy, R.; Turek, V.A.; Kongsuwan, N.; Benz, F.; Carnegie, C.; van de Goor, T.; de Nijs, B.; Demetriadou, A.; Hes Keyser, U.F.; et al. Mapping Nanoscale Hotspots with Single-Molecule Emitters Assembled into Plasmonic Nanocavities U DNA Origami. *Nano Lett.* **2018**, *18*, 405–411. [CrossRef]
59. Kaur, V.; Tanwar, S.; Kaur, G.; Sen, T. DNA-Origami-Based Assembly of Au@Ag Nanostar Dimer Nanoantennas for Label- Sensing of Pyocyanin. *Chemphyschem* **2021**, *22*, 160–167. [CrossRef]
60. Kaur, V.; Sharma, M.; Sen, T. DNA Origami-Templated Bimetallic Nanostar Assemblies for Ultra-Sensitive Detection of Dopa *Front. Chem.* **2021**, *9*, 772267. [CrossRef]
61. Zhu, R.; Li, J.; Lin, L.; Song, J.; Yang, H. Emerging Plasmonic Assemblies Triggered by DNA for Biomedical Applications. *Funct. Mater.* **2020**, *31*, 2005709. [CrossRef]
62. Ding, L.; Liu, B.; Peil, A.; Fan, S.; Chao, J.; Liu, N. DNA-Directed Assembly of Photonic Nanomaterials for Diagnostic Therapeutic Applications. *Adv. Mater.* **2025**, *37*, e2500086. [CrossRef]
63. Ma, W.; Kuang, H.; Xu, L.; Ding, L.; Xu, C.; Wang, L.; Kotov, N.A. Attomolar DNA detection with chiral nanorod assemblies. *Commun.* **2013**, *4*, 2689. [CrossRef]
64. Xu, L.; Gao, Y.; Kuang, H.; Liz-Marzan, L.M.; Xu, C. MicroRNA-Directed Intracellular Self-Assembly of Chiral Nanorod Di *Angew. Chem. Int. Ed.* **2018**, *57*, 10544–10548. [CrossRef] [PubMed]
65. Funck, T.; Nicoli, F.; Kuzyk, A.; Liedl, T. Sensing Picomolar Concentrations of RNA Using Switchable Plasmonic Chirality. *A Chem. Int. Ed.* **2018**, *57*, 13495–13498. [CrossRef] [PubMed]
66. Huang, Y.; Nguyen, M.K.; Natarajan, A.K.; Nguyen, V.H.; Kuzyk, A. A DNA Origami-Based Chiral Plasmonic Sensing De *ACS Appl. Mater. Interfaces* **2018**, *10*, 44221–44225. [CrossRef] [PubMed]
67. Funck, T.; Liedl, T.; Bae, W. Dual Aptamer-Functionalized 3D Plasmonic Metamolecule for Thrombin Sensing. *Appl. Sci.* **20** 3006. [CrossRef]
68. Liu, F.; Li, N.; Shang, Y.; Wang, Y.; Liu, Q.; Ma, Z.; Jiang, Q.; Ding, B. A DNA-Based Plasmonic Nanodevice for Cascade S Amplification. *Angew. Chem. Int. Ed.* **2022**, *61*, e202114706. [CrossRef]
69. Trofymchuk, K.; Glembockyte, V.; Grabenhorst, L.; Steiner, F.; Vietz, C.; Close, C.; Pfeiffer, M.; Richter, L.; Schutte, M.L.; Sel F.; et al. Addressable nanoantennas with cleared hotspots for single-molecule detection on a portable smartphone micros *Nat. Commun.* **2021**, *12*, 950. [CrossRef]
70. Grabenhorst, L.; Sturzenegger, F.; Hasler, M.; Schuler, B.; Tinnefeld, P. Single-Molecule FRET at 10 MHz Count Rates. *J. Am. C Soc.* **2024**, *146*, 3539–3544. [CrossRef]
71. Ochmann, S.E.; Vietz, C.; Trofymchuk, K.; Acuna, G.P.; Lalkens, B.; Tinnefeld, P. Optical Nanoantenna for Single Molecule-B Detection of Zika Virus Nucleic Acids without Molecular Multiplication. *Anal. Chem.* **2017**, *89*, 13000–13007. [CrossRef] [Pub
72. Kaur, V.; Kaur, C.; Sen, T. Single-Molecule Fluorescence Enhancement Based Detection of ATP Using DNA Origami-Asser Au@Ag Nanostar Optical Antennas. *J. Phys. Chem. C* **2023**, *127*, 7308–7318. [CrossRef]
73. Pang, Y.; Wang, C.; Lu, L.; Wang, C.; Sun, Z.; Xiao, R. Dual-SERS biosensor for one-step detection of microRNAs in exosom residual plasma of blood samples for diagnosing pancreatic cancer. *Biosens. Bioelectron.* **2019**, *130*, 204–213. [CrossRef] [Pub
74. Yang, Y.; Song, C.; Zhang, J.; Chao, J.; Luong, H.M.; Zhao, Y.; Wang, L. DNA self-assembled Au nanoparticle clusters on nanorod arrays for high-sensitive and multiplex detection of cancer-related biomarkers. *Nanoscale* **2022**, *14*, 4538–4547. [Cros
75. Zhang, J.; Song, C.; He, X.; Liu, J.; Chao, J.; Wang, L. DNA-mediated precise regulation of SERS hotspots for biosensing bioimaging. *Chem. Soc. Rev.* **2025**, *54*, 5836–5863. [CrossRef]
76. Li, S.; Shi, B.; He, D.; Zhou, H.; Gao, Z. DNA origami-mediated plasmonic dimer nanoantenna-based SERS biosens ultrasensitive determination of trace diethylstilbestrol. *J. Hazard. Mater.* **2023**, *458*, 131874. [CrossRef]
77. Mostafa, A.; Kanehira, Y.; Tapio, K.; Bald, I. From Bulk to Single Molecules: Surface-Enhanced Raman Scattering of Cytoch C Using Plasmonic DNA Origami Nanoantennas. *Nano Lett.* **2024**, *24*, 6916–6923. [CrossRef]
78. Tang, J.; Ji, C.; Lu, X.; Cao, H.; Ling, Y.; Wu, Y.; Qian, L.; He, Y.; Song, B.; Wang, H. DNA Origami Plasmonic Nanoanten Programmable Biosensing of Multiple Cytokines in Cancer Immunotherapy. *Anal. Chem.* **2024**, *96*, 9684–9692. [CrossRef]
79. Sharma, M.; Kaur, C.; Singhmar, P.; Rai, S.; Sen, T. DNA origami-templated gold nanorod dimer nanoantennas: Ena addressable optical hotspots for single cancer biomarker SERS detection. *Nanoscale* **2024**, *16*, 15128–15140. [CrossRef]

https://doi.org/10.3390/molecules310

Kanehira, Y.; Kogikoski, S., Jr.; Titov, E.; Tapio, K.; Mostafa, A.; Bald, I. Watching a Single Enzyme at Work Using Single-Molecule Surface-Enhanced Raman Scattering and DNA Origami-Based Plasmonic Antennas. *ACS Nano* **2024**, *18*, 20191–20200. [CrossRef]
Meng, D.; Ma, W.; Wu, X.; Xu, C.; Kuang, H. DNA-Driven Two-Layer Core-Satellite Gold Nanostructures for Ultrasensitive MicroRNA Detection in Living Cells. *Small* **2020**, *16*, e2000003. [CrossRef]

https://doi.org/10.3390/molecules31030513

molecules

Review

Recent Advances in the Development of Functional Nucleic A… Biosensors Based on Aptamer-Rolling Circle Amplification

Ce Liu [1] and Wanchong He [2,*]

[1] College of Science, Yanbian University, Yanji 133002, China
[2] Shandong Key Laboratory of Applied Technology for Protein and Peptide Drugs, School of Pharmaceutical Sciences and Food Engineering, Liaocheng University, Liaocheng 252000, China
* Correspondence: hewanchong@lcu.edu.cn

Abstract: Aptamers are synthetic nucleic acids or peptides that exhibit high specificity and affinity for target molecules such as small molecules, proteins, or cells. Due to their ability to bind precisely to these targets, aptamers have found widespread use in bioanalytical and diagnostic applications. Rolling circle amplification (RCA) is an amplification technique that utilizes DNA or RNA templates, where circular primers are extended by polymerases to generate multiple repeated sequences, enabling highly sensitive detection of target molecules. The integration of aptamers with RCA offers significant advantages, enhancing both the specificity and sensitivity of detection while ensuring a fast and straightforward process. This synergy has already been widely applied across various fields, including fluorescence, microfluidics, visualization, and electrochemical technologies. Examples include molecular probe development, rapid detection of disease biomarkers, and environmental monitoring. Looking ahead, the aptamer-RCA platform holds great promise for advancing early disease diagnosis, precision medicine, and the development of nanosensors, driving innovation and new applications in these fields.

Keywords: aptamer; rolling circle amplification; biosensor

1. Introduction

Functional nucleic acids (FNAs) are a class of nucleic acid molecules that exhibit a wide range of biological and chemical functions, including molecular recognition, catalysis, gene regulation, and signal amplification. For instance, aptamers function as high-affinity molecular recognition elements, while DNAzymes exhibit catalytic activity similar to that of protein enzymes. Other FNAs, such as ribozymes, regulate gene expression by cleaving target RNAs, and molecular beacons serve as signal transducers in biosensing platforms [1]. Aptamers are a type of functional nucleic acid whose structure is formed by the folding of single-stranded DNA or RNA sequences into complex three-dimensional conformations. This structure endows aptamers with high specificity, high affinity, stability, and reversibility, making them widely applicable in fields such as therapy and biosensing [2]. The ability to chemically synthesize and modify aptamers makes them a powerful alternative to antibodies. Aptamer molecules were first discovered in the 1990s by two independent laboratories, namely the Gold and Szostak groups [3]. Systematic evolution of ligands by exponential enrichment (SELEX) is the core technology for aptamer screening. Its process mainly consists of five key steps: First, construct a nucleic acid library containing a vast number of random sequences (usually 20–100 nucleotides), providing a basis for diversity in screening [4]. Then, incubate this nucleic acid library with target molecules (such as proteins, small molecules, or cells) under optimized conditions to allow

 https://doi.org/10.3390/molecules30…

specific sequences to bind to the targets [5]. Next, separate the nucleic acid molecules that bind to the targets from the unbound ones through methods like solid-phase separation, membrane filtration, or capillary electrophoresis [6]. Subsequently, amplify these bound nucleic acid sequences using PCR (for DNA libraries) or reverse transcription-PCR (for RNA libraries) [7]. Finally, repeat this screening process 8–15 times and gradually increase the stringency of the screening to enrich aptamer sequences with high affinity and high specificity [8]. The development of aptamers, from their discovery to their applications, has progressed rapidly. Advances in SELEX technology and interdisciplinary research have unlocked their immense potential in therapy, biosensing, nanotechnology, and basic research. Despite some challenges, with ongoing technological advancements, aptamers are expected to play an even more significant role in the future of precision medicine and biotechnology [9].

Rolling circle amplification (RCA) is a highly efficient nucleic acid amplification technique that utilizes a circular DNA template to perform isothermal amplification under the action of DNA polymerase, generating a linear single-stranded DNA product containing multiple repeating units. The RCA reaction consists of three core components: a circular DNA template serves as the replication backbone, a primer provides the starting site, and a DNA polymerase (such as Phi29) catalyzes the extension at a constant temperature. After the primer binds to the circular template, the polymerase continuously replicates in a cycle, generating a long-chain DNA containing repeating units. This system can integrate aptamer sequences by designing the template to achieve highly sensitive detection of target molecules, and it is widely used in the field of biosensing [10]. The RCA reaction is characterized by isothermal amplification, high sensitivity, high specificity, and multifunctionality. The characteristics of the RCA reaction are related to the substances that make up the reaction. The material basis of the isothermal reaction is that the temperature at which the reaction occurs is the same as the optimal temperature of the enzyme [11]. The reaction is highly sensitive because it can undergo numerous elongations and amplifications, achieving the corresponding signal amplification in detection. The stability, speed, and sensitivity of the RCA strategy have enabled its widespread application in DNA and protein detection, single nucleotide polymorphism analysis, in situ signal amplification, drug discovery, and nanotechnology [12].

This review introduces the application of RCA combined with aptamers in biosensing detection. Additionally, it points out that this combination offers significant advantages, making it an important tool for highly sensitive and specific detection, and explains that aptamers, as molecular recognition elements, can specifically bind to target molecules (such as proteins [13], small molecules [14], or cells [15]), while the RCA reaction significantly amplifies the binding signal through an isothermal amplification mechanism. This review also shows that the combined strategy not only improves detection sensitivity but also enables accurate detection of trace targets in complex samples, and it further mentions that the versatility of the rolling circle amplification product makes it compatible with a variety of detection methods, such as fluorescence [16], colorimetry [17], and electrochemistry [18], thus making it suitable for fields such as disease diagnosis [19], environmental monitoring and food safety [20]. Furthermore, it predicts that in the future, with the further optimization of aptamer screening technology and the rolling circle amplification reaction system, this strategy will play a greater role in point-of-care testing and precision medicine [21].

2. Aptamer

Aptamers are short, single-stranded DNA or RNA oligonucleotides (typically 15–80 nucleotides) capable of folding into stable, unique tertiary structures. These structures facilitate high-affinity and specific binding to diverse targets through non-covalent interactions such as hydrogen bonding and π–π stacking [22]. Due to their synthetic origin and chemical versatility, aptamers function as effective molecular recognition elements across a broad range of biosensing applications [23].

Their key features—including high specificity [23], stability [24], and ease of synthesis—underpin their growing role in diagnostics and therapeutics. Unlike antibodies, aptamers can be chemically synthesized and modified, which allows for scalable, cost-effective production with high batch-to-batch consistency [24]. The conventional SELEX method is widely used for aptamer selection but is often time-consuming. Recent optimization strategies have significantly improved screening efficiency. For instance, Yang et al. used capillary electrophoresis-SELEX (CE-SELEX) to isolate a high-affinity aptamer for human apotransferrin within only three rounds [25]. Similarly, Liu et al. developed a microfluidic-based protein microarray system–SELEX (PMM-SELEX), achieving reproducible aptamer selection with high throughput and precision [26].

Aptamers are now being developed for use in neurological and cancer therapies, pathogen diagnostics, and precision drug delivery [27]. Nonetheless, challenges such as rapid renal clearance, structural instability, and potential off-target effects hinder their full therapeutic potential. Ongoing research in chemical modification and delivery systems is vital to overcoming these limitations and enhancing their applicability in clinical settings [28].

3. The Regulation of RCA Reaction

By regulating the accessibility of at least one of the four essential components of RCA, the initiation of the RCA reaction can be controlled. These four essential components are as follows: the primer, the circular DNA template (CDT), the DNA polymerase (usually φ29 DNA polymerase unless otherwise specified), and deoxyribonucleoside triphosphates (dNTPs). To date, there have been no reported methods for regulating the RCA reaction by modulating the accessibility of the DNA polymerase or dNTPs. Here, we focus on the techniques specifically designed to regulate the accessibility of the primer strand or the circular DNA template.

3.1. *The Design of Primers*

3.1.1. The Primer Is Directly Connected to the Aptamer

In an early example, some scientists utilized a thrombin-binding structure-switching aptamer to regulate the RCA reaction through target-induced primer displacement [29]. In this case, the primer was first captured by an immobilized DNA probe at its 5′ end. After performing a washing step, the CDT and φ29 DNA polymerase were added to the captured primer, enabling the generation of target-mediated rolling circle amplification products from the 3′ end.

Over the past few years, Bialy, RM et al. designed a method that combines the primer and the aptamer without the need for an immobilized DNA probe or a washing step [30]. In this approach, the 5′-end of the primer is hybridized with the circular DNA template, and the other 3′-end is bound to the aptamer. In the absence of the target substance, the aptamer prevents the primer from binding to the template; thus, the rolling circle amplification reaction cannot be initiated. When the target substance is added to the system, the aptamer is released and binds to the target substance. The 3′ end of the primer successfully binds to the circular template, forming a complementary primer-template double strand [31]. Under the catalytic action of φ29 DNA polymerase, the RCA reaction proceeds smoothly (Figure 1A).

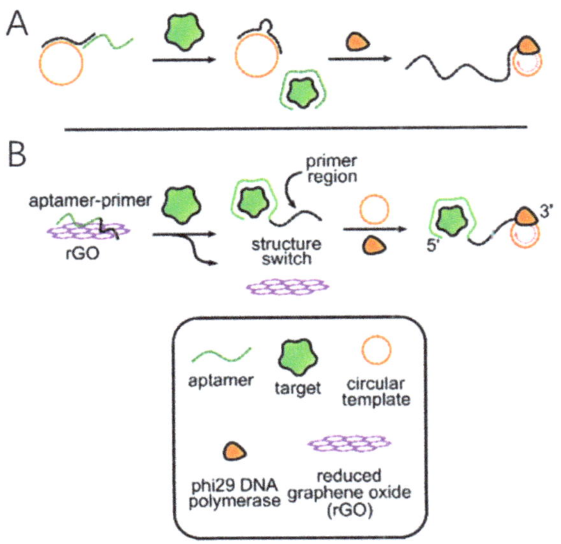

Figure 1. Structure switching primer design strategies: (**A**) Tripartite structure-switching. (**B**) Structure switching using reduced graphene oxide material [30].

However, to ensure the smooth progress of the RCA reaction, the aptamer must be smoothly released upon target binding, and at the same time, the primer should have a certain affinity for the circular DNA template. To achieve these two points, it is important to introduce a new substance that can both immobilize the aptamer and make it easily detachable. In 2009, scientists proposed using graphene oxide to adsorb the primer–aptamer complex. Graphene oxide has adjustable properties that enable it to retain single-stranded or double-stranded DNA on its surface. When the aptamer binds to the target substance to form a complex, the aptamer adsorbed on the surface of the graphene oxide will detach, thus initiating the rolling circle amplification reaction (Figure 1B). According to a reported multifunctional biosensing platform, reduced graphene oxide can non-specifically adsorb single-stranded DNA molecules, and DNA aptamers can bind to reduced graphene oxide. The effect of reduced graphene oxide in the RCA reaction is better than that of graphene oxide [32].

3.1.2. RNA-Cleaving DNAzyme (RCD) Mediated Primer Regulation

RCD is a class of functional DNA molecules with RNA cleavage activity. It can specifically recognize and cleave RNA substrates just like protein ribonuclease [33] and has important applications in the fields of biotechnology and medicine. It can be used in combination with the RCA reaction. In the RCA reaction, RCD can cleave the RNA sequences in the RCA products to release reporter molecules. It can also initiate downstream reactions to release a new set of RCA primers, enabling multiple rounds of signal amplification. Several strategies have been developed to regulate primers in the RCA reaction mediated by RCD.

(1) RCD Cleavage for Primer Release

In the RCA reaction, the primer precursor can be designed to contain the specific cleavage site of RCD (such as an RNA sequence or an RNA-DNA hybrid strand). In the initial state, the primer is in a closed or inactive state. When the target (such as a specific RNA or metal ion) activates the RCD, the RCD cleaves the primer precursor, releasing an active primer with a 3′-OH end, thus initiating the RCA amplification. This method is highly target-dependent and can effectively avoid non-specific amplification, making it suitable for the highly sensitive detection of low-abundance RNAs (such as miRNA or viral RNA). For example, in nucleic acid detection, after the target RNA activates the RCD, it cleaves the primer precursor to release the functional primer, which then binds to the circular template for RCA, achieving signal amplification.

(2) RCD Regulation of Primer Conformation

By designing the primer as a hairpin structure or a molecular beacon and embedding the cleavage site of RCD (such as an RNA sequence) in its stem, the dynamic regulation of primer activity can be achieved. In the unactivated state, the primer cannot bind to the template due to the closure of its secondary structure. When the RCD is activated by the target, it cleaves the hairpin stem, causing the primer conformation to change from a closed state to a linear structure, exposing the sequence complementary to the circular template and thus initiating the RCA. This regulation method can significantly reduce the background signal and improve the signal-to-noise ratio of the detection, especially suitable for RNA imaging in living cells or real-time monitoring. For example, in fluorescence detection, after the RCD cleaves the hairpin primer, the fluorescent group and the quenching group are separated, and at the same time, the active primer is released to trigger the RCA, achieving double signal amplification.

(3) RCD-Mediated Primer Generation

The RCD can directly generate new RCA primers by cleaving long-chain RNAs or DNA-RNA hybrid strands instead of relying on pre-designed primer precursors. When the target is present, the RCD recognizes and cleaves a specific sequence, and the generated short fragment (such as a DNA containing a 3′-OH end) can serve as a functional primer to initiate the RCA. This method is suitable for complex samples (such as cell lysates or clinical samples) and can achieve exponential signal amplification through cascade reactions. For example, in the detection of viral RNA, the target hybridizes with the auxiliary probe to form the cleavage site of the RCD. After the RCD cleaves it, a short DNA primer is generated, which in turn triggers the RCA amplification. Combined with fluorescent or electrochemical signal output, highly sensitive detection can be achieved.

(4) RCD Regulation of Primer Competition

By introducing competitive primers (containing the cleavage site of RCD) to coexist with functional primers, the dynamic regulation of RCA can be achieved. The competitive primers preferentially bind to the circular template but do not extend, thus inhibiting the function of the functional primers. When the RCD is activated by the target, it cleaves the competitive primers to make them dissociate, release the closure of the template, and enable the functional primers to dominate the RCA reaction. This strategy can be used to construct logical gating systems (such as AND/OR gates) to achieve the coordinated detection of multiple targets. For example, in nucleic acid computing, two targets, respectively, activate different RCDs. After cleaving the corresponding competitive primers, the functional primers initiate the RCA and output specific signals.

(5) RCD Control of Primer Recycling

The RCD can continuously cleave the primer complementary sequence in the RCA extension product, releasing the bound primer and allowing it to participate in the amplification cycle again. This method can significantly improve the utilization rate of primers and achieve continuous signal amplification, which is suitable for the detection of ultra-low concentration targets. For example, multiple RCD cleavage sites are embedded in the long-chain RCA product. The RCD repeatedly cleaves to release the primer, initiating multiple rounds of amplification. Combined with fluorescent or nanoparticle signal output, the detection sensitivity can be improved to the single-molecule level.

3.2. The Design of Circular Template

3.2.1. Ligate the Circular Template

The principle of constructing circular templates mediated by T4 DNA ligase is based on the enzymes' ability to catalyze the formation of a phosphodiester bond between the 5′-phosphate group and the 3′-hydroxyl group of DNA molecules, thereby achieving the circular ligation of linear DNA molecules [34]. This method is applicable to both single-stranded and double-stranded DNA templates. For single-stranded DNA, the 5′ end needs to be phosphorylated in advance, while double-stranded DNA needs to be denatured into single-strands before use. There are two types of ligation methods: high-efficiency (>90%) sticky-end ligation and low-efficiency (about 20%) blunt-end ligation. The former requires the design of complementary overhanging end sequences. The characteristics of this technology include mild reaction conditions (usually carried out at 16 °C) and good stability of the ligation products. It is particularly suitable for the circularization of DNA in the length range of 50–1000 nt. However, it has limitations, such as low blunt-end ligation efficiency and the tendency to produce multimer by-products. The circularization efficiency can be improved by optimizing reaction conditions (e.g., adding PEG8000 and controlling DNA concentration).

3.2.2. Design Circular Aptamers

Circular aptamers play multiple crucial roles in the RCA reaction. Their unique structural characteristics significantly enhance the amplification efficiency and detection performance [35]. As a core component of the circular template, circular aptamers optimize the RCA reaction through the following mechanisms: First, their closed-loop structure eliminates the end-inhibition effect, enabling DNA polymerase to perform continuous and efficient strand-displacement synthesis, which boosts the amplification efficiency by 10–100 times. Second, the aptamer sequence can specifically bind to target molecules (such as proteins, small molecules, or cell markers). In the reaction, it serves as both a recognition element and a signal transducer, achieving integrated "recognition–amplification" detection. For example, in thrombin detection, the circular aptamer can specifically bind to the target protein and also act as an RCA template to guide the synthesis of repetitive sequences [36]. Through cascade amplification, a single binding event is converted into thousands of copies of nucleic acid signals. In addition, circular aptamers can integrate functional sequences (such as promoters, ribozyme sites, or fluorescent labels) through modular design to achieve multiplex detection or intelligent regulation. Compared with linear aptamers, the circular structure also significantly improves the resistance to nucleases, allowing the reaction to remain stable in complex biological samples. These characteristics have led to the widespread application of RCA technology based on circular aptamers in fields such as ultrasensitive biosensing, single-cell analysis, and in vitro diagnostics, with a detection limit reaching the fM or even aM level. The latest research has also developed dynamic response systems through engineered circular aptamers, such as allosteric aptamers sensitive to pH or temperature, further expanding the application prospects of RCA in in vivo monitoring and intelligent diagnosis.

The inherent circular structure of aptamers also enables their integration into intramolecular structure-switching systems for regulating rolling products. Zhao et al. transformed a linear aptamer targeting lipopolysaccharides into an aptamer by incorporating the aptamer and a primer-binding region within a CDT. In the absence of the target, the aptamer formed a dumbbell-shaped structure through intramolecular hybridization, which restricted access to the primer-binding region. When lipopolysaccharides were introduced, the aptamer underwent a conformational change, adopting an extended shape that exposed the primer-binding region, thereby facilitating primer binding and rolling product generation in a single-pot assay.

Roger's research team developed a structural switching system using circular aptamers, where the binding of platelet-derived growth factor induces the detachment of the circular platelet-derived growth factor aptamer from reduced graphene oxide [30]. Once detached, the complex formed between the circular aptamer and platelet-derived growth factor is separated from reduced graphene oxide via centrifugation. The supernatant containing the released circular aptamer is then mixed with primers, enabling the generation of RCA products. A limitation of this method is the need for a centrifugation step to remove the reduced graphene oxide, as reduced graphene oxide can adsorb linear primers, thereby hindering the RCA reaction.

Another distinctive approach involves incubating a circular aptamer for glutamate dehydrogenase with magnetic beads coated with recombinant glutamate dehydrogenase to form a circular aptamer–recombinant glutamate dehydrogenase complex [30]. The addition of natural glutamate dehydrogenase causes the circular aptamer to be released from the magnetic beads. After removing the beads, the released circular aptamer can bind to the added primers, enabling an RCA reaction mediated by the natural glutamate dehydrogenase.

Most of the examples mentioned above require a separation step, which presents a significant challenge for their implementation in point-of-care detection. To overcome this issue, a method has been developed that utilizes a CDT integrated with an aptamer to regulate the RCA reaction by obstructing the ability of Phi29 DNA polymerase to transcribe the circular DNA template upon target binding. When a platelet-derived growth factor aptamer is integrated into the circular DNA template, it was observed that upon binding of the target protein, Phi29 DNA polymerase is unable to displace the bound protein, thus preventing it from transcribing the aptamer sequence and inhibiting the RCA reaction [37]. This approach eliminates the need for a separation step, simplifying the detection process considerably.

4. The Amplification Methods of the RCA Reaction

In RCA technology, there are two primary amplification modes—linear and exponential-distinguished by their underlying amplification mechanisms and efficiencies. These two modes exhibit significant differences in both reaction pathways and applicable scenarios.

4.1. Linear RCA

Linear RCA is the most basic form of amplification, where the core mechanism involves continuous rolling synthesis by a DNA polymerase on a circular template, generating long single-stranded DNA products containing hundreds to thousands of repeat units (Figure 2A). This process is characterized by three typical features: (1) each circular template can only produce a single continuous multi-copy product, with amplification efficiency linearly correlated with reaction time; (2) no additional primers or enzymes are required, making the reaction system simple; (3) the typical signal amplification factor ranges from 10^3 to 10^4. Linear RCA is particularly suited for scenarios that require stable and controlled amplification, such as in situ hybridization and protein detection. For instance, in immuno RCA technology, linear amplification is initiated by primers conjugated to antibodies, enabling the visual detection of individual antigen molecules.

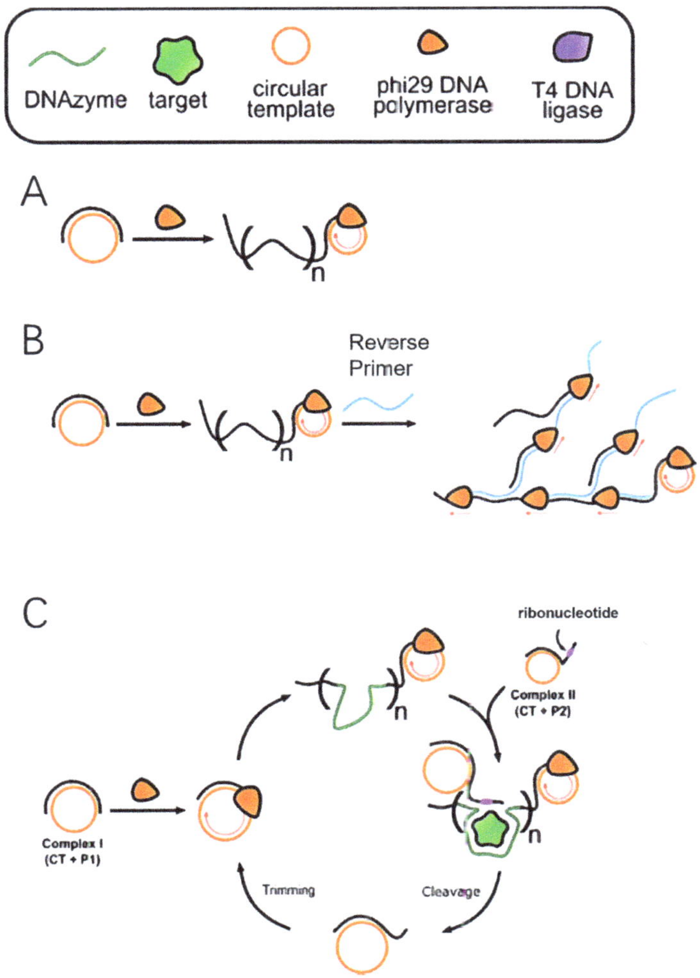

Figure 2. Schematic illustrations for linear and exponential amplification of RCA: (**A**) Linear RCA. (**B**) Hyperbranched RCA. (**C**) DNAzyme feedback amplification [30].

4.2. Exponential RCA

Exponential RCA achieves signal amplification through the introduction of a multistage amplification mechanism, which can be implemented in two main ways:

(1) Hyperbranched RCA: Reverse primer sites are designed on the initial RCA products, and cascade amplification mediated by multiple primers leads to the formation of a three-dimensional branched structure (Figure 2B). This method can increase the amplification factor to 10^6–10^9 times, but precise control of primer concentration is necessary to avoid non-specific amplification [38].

(2) Primer regeneration RCA[38]: Nucleic acid endonucleases, such as Nickase or RCD, cyclically cleave RCA products, releasing new primers that initiate secondary amplification cycles (Figure 2C). This dynamic regulation approach is particularly useful for real-time quantitative detection, such as the ultrasensitive analysis of viral nucleic acids.

5. Applications of the Combination of Aptamers and RCA

The integration of aptamers with RCA technology has become a prominent approach in molecular diagnostics and biosensing. This strategy is particularly valuable due to its ability to combine the exceptional specificity of aptamers with the enhanced signal amplification offered by RCA. Key applications of this synergistic combination encompass (1) fluorescence-based detection, (2) microfluidic systems, (3) imaging methods, and (4) electrochemical sensing techniques.

5.1. Fluorescence-Based Detection

Zhu et al. designed an aptasensor utilizing a novel form of exponential rolling circle amplification for the label-free and highly sensitive fluorescence detection of ochratoxin A (OTA) [16]. G-quadruplex (G4) refers to a well-ordered, four-stranded DNA structure that forms from a single-stranded, G-rich nucleic acid sequence facilitated by the presence of K^+ and Na^+ [39].

The approach involved a circular amplification template and a multifunctional double-stranded probe, aptamer probe–hairpin primer probe (APH). APH was formed by hybridizing an aptamer probe with a hairpin primer probe in a 1:1 ratio. The aptamer probe, consisting of 51 nucleotides, included a 36-nucleotide aptamer sequence at the 3′ end that specifically recognized OTA and a 27-nucleotide region at the 5′ end designed to hybridize with RCA amplification products. The hairpin primer probe, with a stem length of 9 nucleotides, contained a 24-nucleotide loop region that was complementary to the central part of the aptamer probe, along with a 12-nucleotide primer sequence split into two segments: nine nucleotides at the 5′ end and three nucleotides at the 3′ end. These sequences were responsible for recognizing and hybridizing with the circle template, thereby triggering the RCA reaction. In the absence of OTA, the APH probe remained stable, preventing the initiation of RCA. However, when OTA was present, it bound tightly to the aptamer region, leading to the release of the hairpin probe. This release induced a conformational change in the hairpin probe, transforming it into a hairpin structure, which then hybridized with the circle template, initiating RCA. The RCA process produced amplicons containing repetitive G4 dimer structures and antisense feedback sequences. The antisense sequences interacted with the APH complex, replacing the old hairpin probe with a new one via toehold strand displacement, thus enabling the continuous initiation of RCA and the exponential amplification of the products. Simultaneously, the G4 structures formed on the RCA amplicons interacted with thioflavin T, resulting in significantly enhanced fluorescence due to the high quantum yield of the fluorescence signal (Figure 3A). This system allowed for label-free, highly sensitive detection of OTA by measuring the fluorescence intensity of the G4 dimer/thioflavin T complex [16].

Wang et al. developed a sandwich-type detection strategy that combines antibody-conjugated magnetic beads and aptamers, utilizing RCA and the nicked PAM/CRISPR-Cas12a system as signal transduction mechanisms for highly sensitive C-reactive protein detection (Figure 3B). Upon the introduction of C-reactive protein, the aptamer–primer probe bound to C-reactive protein and was subsequently captured by the antibody-conjugated magnetic beads, forming a sandwich-like complex. This complex then triggered the RCA reaction, amplifying the signal. As a result, the RCA products activated CRISPR-Cas12a trans-cleavage activity, which in turn generated a fluorescent signal [40].

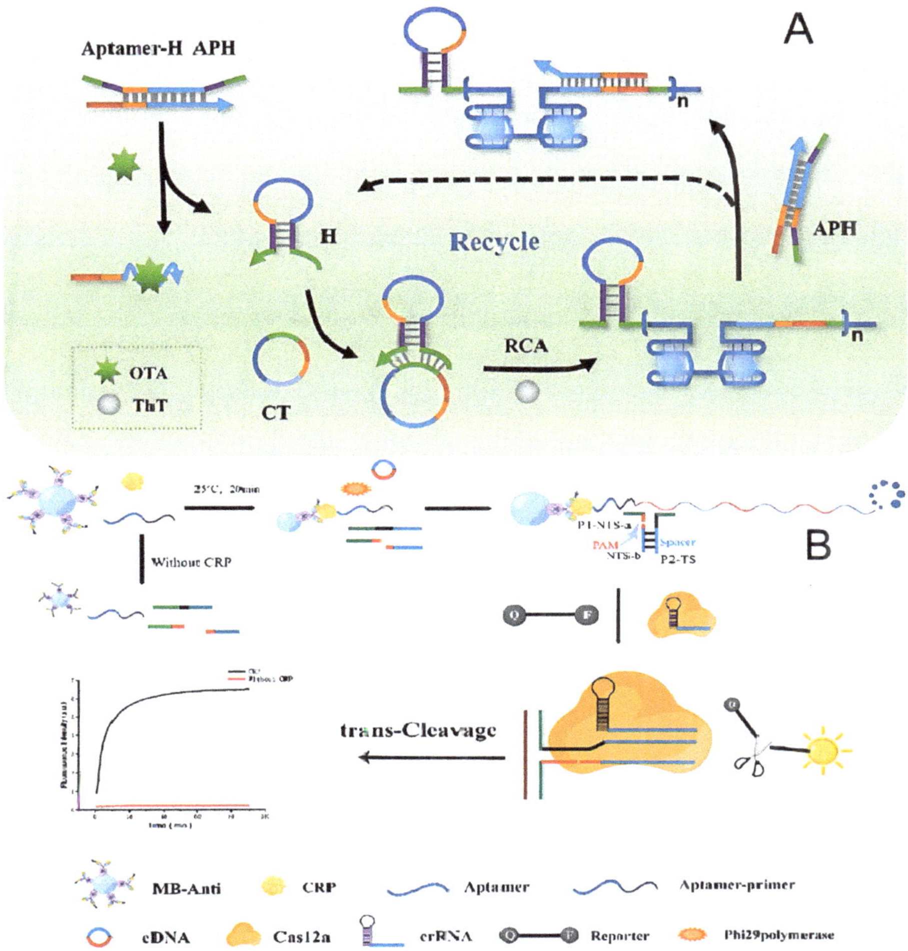

Figure 3. (**A**) Illustration of the fabrication of an aptasensor based on an RCA fluorescence biosensor for OTA detection. H probe was released in the presence of OTA to combine with CT to trigger RCA reaction to produce a large number of G4 signal structures (ochratoxin A (OTA), hairpin (H), circle template (CT), Thioflavin T (ThT)) [16]. (**B**) The principle of C-reactive protein assay based on RCA and CRISPR-Cas12a system dual signal amplification strategy (magnetic bead–antibody (MB-Anti), C-reactive protein (CRP)) [40].

Niazi et al. developed a fluorescent aptasensor based on RCA and the fluorescence quenching effect of g-C_3N_4 nanosheets for the detection of aflatoxin M1. The aptamer, serving as a primer, initiates RCA in the presence of the target. In the absence of aflatoxin M1, the RCA template hybridizes with the aptamer, forming a circular DNA that undergoes circularization via T4 DNA ligase. This circular DNA then triggers RCA, facilitated by phi29 DNA polymerase, producing a long ssDNA product with multiple copies of the target aptamer sequence. The RCA product is labeled with complementary DNA, which acts as a signal probe for target detection. This circular DNA binds to multiple sites on the RCA product to form a duplex with time-resolved fluorescence nanoparticles (TRFNPs). When no target is present, the RCA product/TRFNPs-circular DNA duplex cannot bind to the g-C_3N_4 nanosheet due to the stronger affinity of g-C_3N_4 for ssDNA over dsDNA, meaning the fluorescence of the probe remains unquenched (signal on). However, when aflatoxin M1 is present, the aptamer preferentially binds to the target, inhibiting the hybridization of the aptamer with RCT and halting RCA. As a result, no RCA product is formed, and the free circle DNA is adsorbed onto g-C_3N_4, leading to fluorescence quenching. The extent of fluorescence quenching increases with the target concentration, as more aptamers bind to aflatoxin M1, preventing hybridization with RCA product and increasing the amount of free circle DNA (Figure 4). The strong interaction between g-C_3N_4 and ssDNA quenches the fluorescence via static quenching, allowing for quantitative measurement of the target by monitoring the fluorescence intensity after adding g-C_3N_4 [41].

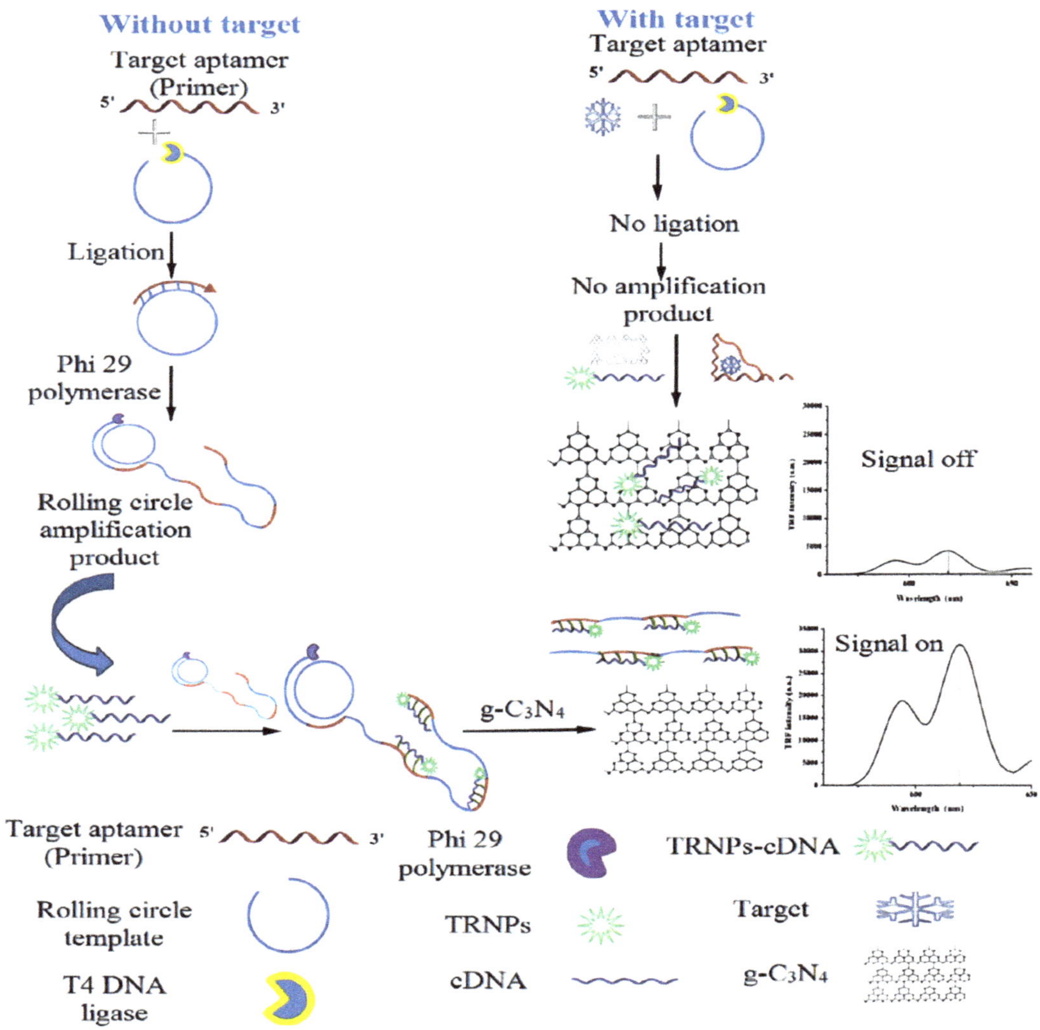

Figure 4. Schematic design of aflatoxin M1 detection using g-C_3N_4 nanosheet and time-resolved fluorescence nanoparticles combined with RCA-based DNA amplification. In the absence of aflatoxin M1, the RCA reaction was triggered by target aptamer (primer) to dismantle cDNA on the g-C_3N_4 surface and produce a fluorescence signal finally (time-resolved fluorescence nanoparticles (TRFNPs)) [41].

Zhang et al. developed a method where streptavidin-coated magnetic beads were functionalized with an aptamer targeting the receptor-binding domain through a biotin-tagged complementary DNA strand (biotin-cDNA). Upon binding to the receptor-binding domain, the aptamer was released from the biotin-cDNA, allowing the cDNA to initiate RCA on the surface of the magnetic beads. The detection of the receptor-binding domain was accomplished through a dual-signal approach. For fluorescence detection, the RCA products were combined with a dsDNA probe, which was labeled with both a fluorophore and a quencher. The hybridization of RCA products with the dsDNA probe resulted in the separation of the fluorophore and quencher, leading to fluorescence emission (λex = 488 nm, λem = 520 nm) [42].

A method for detecting multiple pathogenic microorganisms is described, utilizing a DNA composite system encapsulating DNA-stabilized silver nanoclusters (AgNCs) and graphene oxide in conjunction with RCA. Initially, two distinct RCA-based DNA composites are constructed, each linked to a probe consisting of DNA-stabilized AgNCs and an ssDNA aptamer specific to two different bacterial pathogens. Graphene oxide is then incorporated into the system to capture the ssDNA aptamers from the DNA composites and act as a selective quencher for the fluorescence of the DNA/AgNCs complex. When the target bacteria are recognized, the ssDNA aptamer binds to the bacteria, causing it to detach from the graphene oxide surface. As a result, the RCA-based DNA composite generates a strong fluorescent signal [43].

Roger et al. present a streamlined approach to integrating protein-binding aptamers with isothermal amplification, enabling a one-step reaction that does not require any labeled or modified DNA species. This method is easily adaptable for use in paper-based point-of-care devices. The assay employs a protein-binding aptamer that also functions as a linear primer, initiating RCA, referred to as an "aptaprimer". In the absence of a target, the aptaprimer binds to a complementary circle template and triggers RCA in the presence of phi29 DNA polymerase and dNTPs, producing a long DNA product. This product can bind fluorescent dyes like SYBR Gold™ or QuantiFluor™ to generate a fluorescence signal (Figure 5A). When a protein target is introduced, the aptaprimer forms an aptamer–protein complex, which prevents the aptaprimer from binding to the circular template, thereby inhibiting the RCA process and modulating the fluorescence output. In the AP, the primer-binding region is incorporated within the native aptamer at the 3′ end. Polymerization of the AP strand occurs at the 3′ end, and the hybridized AP is displaced by f29DP as it proceeds through the circular template (Figure 5B). The scientists demonstrate the application of this system for detecting two proteins—platelet-derived growth factor and thrombin—showcasing the potential to couple protein-mediated isothermal DNA amplification with a detectable output signal in a simple, one-step assay format. This system can be employed either in a solution or as part of a paper-based point-of-care test (Figure 5C) [37].

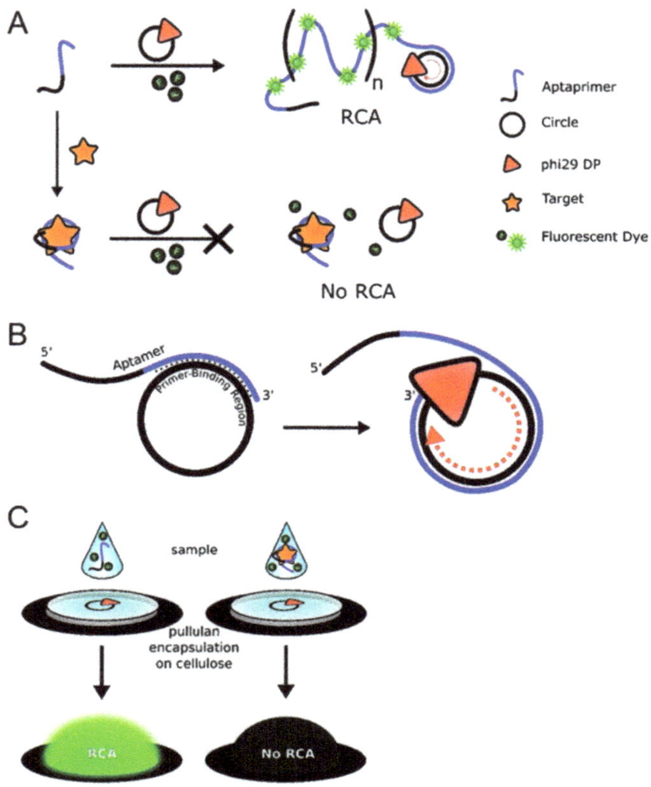

Figure 5. (**A**) Schematic representation of an aptaprimer for target-mediated generation of RCA product by incorporating an intercalating dye such as SYBR Gold or QuantiFluor. The addition of a protein target prevents the binding of the aptaprimer to the circular template, inhibiting RCA. (**B**) Aptaprimer concept: the circle-binding region is embedded within the native aptamer strand on the 3′-end. Polymerization occurs from the 3′-end with phi29DP displacing hybridized aptaprimer to continue elongation of the RCA product. (**C**) A sample mixed with aptaprimer and intercalating dye is spotted onto a wax-contained cellulose well holding pullulan-encapsulated RCA reagents (phi 29DP, circle template, and dNTPs), allowing for real-time one-step RCA and production of a fluorescence signal [37].

5.2. Microfluidic Systems

In recent years, aptamer-based microfluidic platforms have seen rapid development, with a growing focus on improving their detection sensitivity. A study presents an innovative RCA approach for highly sensitive whole-cell detection using microfluidic devices [44]. The dual-RCA method comprises two components: a capturing RCA reaction and a signaling RCA reaction [45]. The capturing RCA reaction involves modifying the microfluidic channel surfaces with long, tandem repeating aptamers (poly-aptamers), which effectively capture target *E. coli* O157:H7 cells. Scientists show that poly-aptamer-modified microchannels capture three times as many target cells compared to those modified with mono-aptamers.

In addition, the signaling RCA method is integrated into the dual-RCA approach to significantly amplify the detection signal. The findings reveal that the signal is enhanced by up to 50 times when signaling RCA is used, compared to traditional single fluorescence probes. When both capturing RCA and signaling RCA are combined in one system, the detection signal is further amplified by approximately 250 times. In conclusion, this microfluidic platform, combined with the dual-RCA approach, offers a simple yet promising solution for sensitive whole-cell detection, making it highly suitable for food safety testing [44].

In the study by Yu et al., a dual-mode aptasensor was developed that combines colorimetric detection with a microfluidic chip to achieve the desired outcomes [46]. This aptasensor enables rapid on-site screening that can be visually assessed while also allowing for the simultaneous quantification of multiple bacterial species. Specifically, the presence of pathogenic bacteria, such as Salmonella typhimurium (S.T.) and Vibrio parahaemolyticus (V.P.), was initially detected by visual inspection. Subsequently, the microfluidic chip was used to quantify the levels of S.T. and V.P. in positive samples. To generate the necessary detection signals, a set of magnetic DNA-encoded probes was created (Figure 6A). These probes contained RCA-produced long DNA strands enriched with G-quadruplex sequences. These sequences, in conjunction with hemin, function as DNAzymes that catalyze a reaction with the 3,3′-5,5′-Tetramethylbenzidine-H_2O_2 system, producing a colorimetric change. Additionally, the probes were cleaved by EcoRV endonuclease to generate DNA fragments of varying lengths. The microfluidic chip was then used to separate and quantify these DNA fragments, enabling the simultaneous quantification of S.T and V.P (Figure 6B). Using this method, as few as 100 CFU/mL of either V.P. or S.T. could be detected visually, while the microfluidic chip allowed detection down to 32 CFU/mL for S.T. and 30 CFU/mL for V.P. within just 3 min. This dual-mode aptasensor proved to be highly efficient, offering both rapid screening and precise quantification of bacterial contamination in food samples, showcasing its potential for on-site detection and simultaneous quantification of foodborne pathogens.

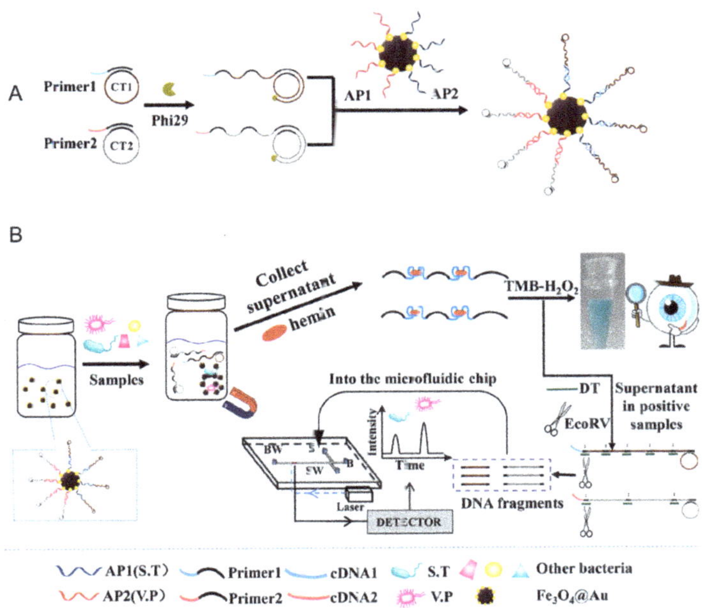

Figure 6. (**A**) The illustration of the preparation of magnetic DNA-encoded probes. RCA produces were linked on magnetic bead surface (circle template 1, 2 (CT1, CT2), 3,3′,5,5′-Tetramethylbenzidine-hydrogen peroxide (TMB-H_2O_2)). (**B**) The dual-mode aptasensor for judging the presence and simultaneous determination of S.T and V.P based on RCA reaction in a microfluidic chip. In the presence of S.T. and V.P., they are captured by magnetic DNA-encoded probes to produce a visual signal (target DNA (DT), *Escherichia coli* RY13 V (EcoRV)) [46].

5.3. Visual Detection

Gao et al. developed a dual-mode aptasensor integrating both colorimetric and electrochemical detection, utilizing G4 produced through RCA [47]. During the RCA process, specific binding between the aptamer and target led to the generation of a large number of G4. The colorimetric component of the sensor was based on the interaction between the G4 and hemin, which modulated the 3,3′,5,5′-Tetramethylbenzidine reaction, enabling visual and semi-quantitative detection of kanamycin. For electrochemical detection, the G4 strongly interacted with methylene blue, facilitating the generation of an electrical signal.

This dual-model approach combined the simplicity of colorimetric visualization with the high sensitivity and precision of electrochemical measurements. The aptasensor demonstrated excellent specificity, minimizing potential interference. Additionally, it was effectively applied to monitor kanamycin levels in milk, showing great potential for practical use in food safety and monitoring applications.

In the study by Wang et al., a visual nanoplatform was developed for highly sensitive and precise detection of glutathione using a dumbbell DNA-mediated RCA method [48]. The system consists of a nicked dumbbell probe (H1/H2) and a single-stranded probe (L0). Upon the addition of glutathione, the disulfide bond in the L0 probe is cleaved, producing L1 and L2. These fragments then hybridize with the loop region of H1, initiating the circularization of the dumbbell probe, which serves as a template for RCA (Figure 7A). This triggers the amplification process, during which horseradish peroxidase is encapsulated in three-dimensional DNA flower structures using a one-step method. By converting the detection of unstable glutathione into stable horseradish peroxidase–DNA flower structures with enhanced enzyme activity, this biosensor demonstrates ultra-high sensitivity and accuracy. It successfully detects glutathione in various cancer cell lines, showcasing its potential for precise biomarker analysis.

Zhang et al. developed a highly sensitive and specific visual detection method for aflatoxin B1 that leverages the target selectivity of aptamers, RCA, and enzyme-catalyzed biological amplification [49]. In this approach, the aflatoxin B1-specific aptamer is immobilized onto the surface of magnetic beads, serving as the molecular recognition element. In the absence of aflatoxin B1, the aptamer and the auxiliary linker probe remain in a double-stranded configuration due to partial base pairing. However, in the presence of aflatoxin B1, the aptamer binds selectively to aflatoxin B1, causing the linker probe to return to a single-stranded form. This triggers the RCA process, where the single-stranded linker probe acts as a template to generate long DNA strands. These strands subsequently capture large quantities of signal probes and horseradish peroxidase enzymes. The horseradish peroxidase enzymes then catalyze the oxidation of 3,3′,5,5′-Tetramethylbenzidine by hydrogen peroxide, resulting in a color change from colorless to deep blue, providing a visual signal. This method achieves high sensitivity and specificity for aflatoxin B1 detection (Figure 7B). Furthermore, the aptasensor exhibited remarkable selectivity for aflatoxin B1 when compared to five other common mycotoxins. Notably, all reactions take place on the surface of the magnetic beads, simplifying the detection process. This setup enhances the efficient isolation and collection of aflatoxin B1 from complex sample matrices, improving the sensor's selectivity and resistance to interference, highlighting its potential for practical application in real-world aflatoxin B1 detection.

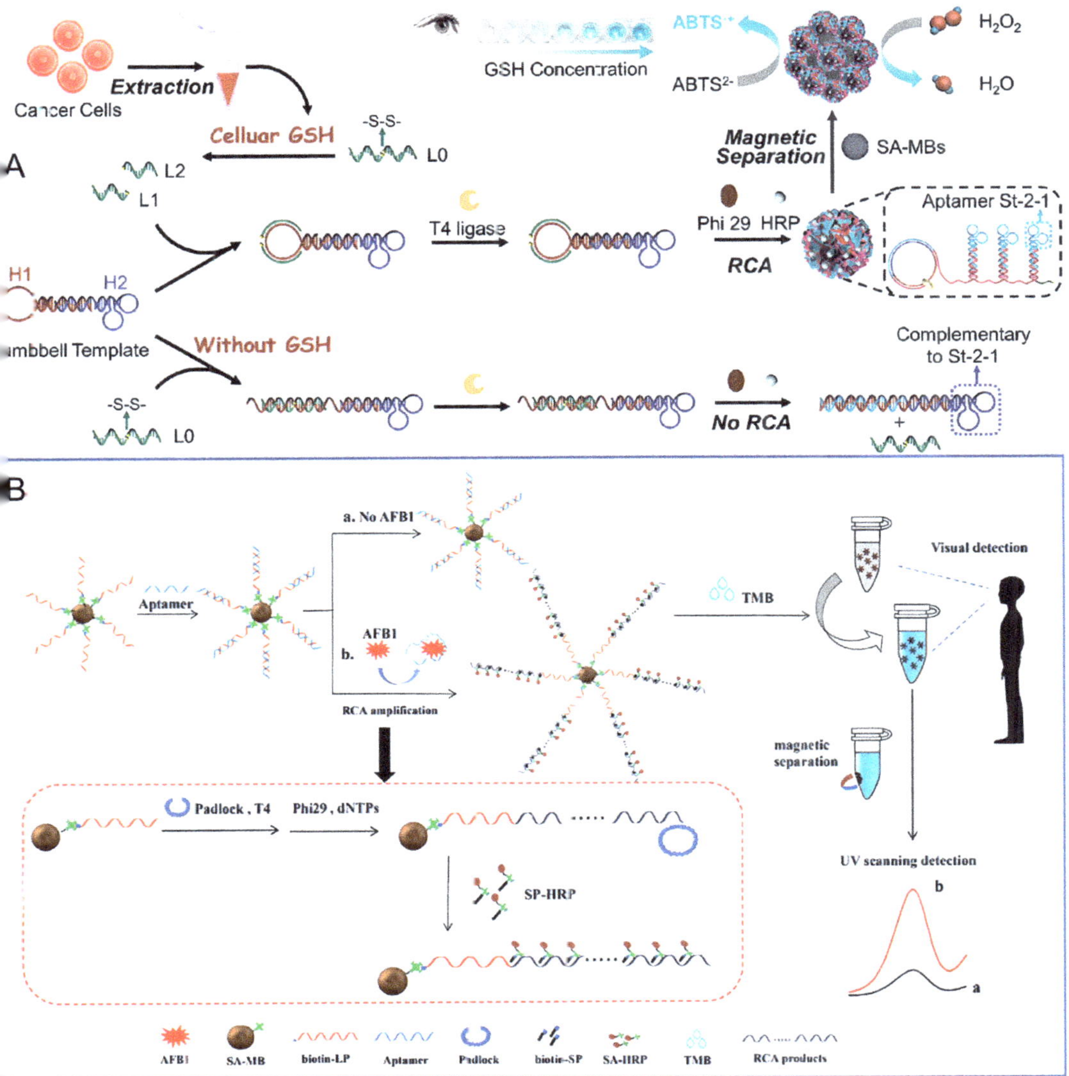

Figure 7. (**A**) Schematic illustration of disulfide cleavage-induced visual biosensing of glutathione based on the dumbbell DNA-mediated RCA for self-assembly of horseradish peroxidase–DNA flower structures using magnetic separation to improve sensitivity (glutathione (GSH), horseradish peroxidase (HRP), streptavidin-magnetic bead (SA-MB), 2, 2′-azino-bis(3-ethylbenzothiazoline-6-sulfonic acid (ABTS)) [48]. (**B**) Schematic illustration of the aflatoxin B1 detection principle based on the aptamer capture and triggered RCA reaction. In the presence of aflatoxin B1, the RCA reaction was triggered and produced a large number of repeated products that can combine with HRP to produce visual signals [49].

Yurdusev group used a model aptamer binding SARS-CoV-2 Spike glycoprotein as a template, triggering the RCA system in a ligand-dependent manner. They confirmed the presence of the Spike protein in the test samples at concentrations within the nanomolar range using an adapted design derived from this model aptamer-RCA system [50].

5.4. Electrochemical Sensing Techniques

The accurate quantification of tumor-derived exosomes, which are emerging as highly promising non-invasive biomarkers for cancer diagnosis, is crucial. In response to this need, a novel bispecific-aptamer-based sandwich-type electrochemical aptasensor (Figure 8) was developed [51]. This sensor utilizes gold nanoparticles and incorporates a four-way junction-induced dual RCA-assisted strategy for highly specific and sensitive detection of exosomes. In this design, the aptasensor employs two specific aptamers: one targeting CD63 and the other recognizing the cancer-related protein mucin-1. The CD63 aptamer is immobilized on a gold electrode to capture exosomes, forming the initial complex. When the mucin-1 aptamer is introduced, it binds and completes the sandwich structure. The 3′ end of the MUC1 aptamer facilitates the formation of the four-way junction, aided by a molecular beacon probe and a binary DNA probe. This assembly triggers a dual-RCA reaction, which is initiated when two cytosine-rich circular DNA templates bind to the ends of the four-way junction. The RCA process generates dual-amplified products that contain multiple G4 structures, which in turn trap methylene blue indicators and significantly enhance electrochemical signals. The resulting aptasensor demonstrated exceptional specificity, sensitivity, reproducibility, and stability in detecting exosomes derived from MCF-7 cells. Moreover, this sensor showed excellent potential for use in clinical applications, effectively recovering exosomes from normal human serum samples.

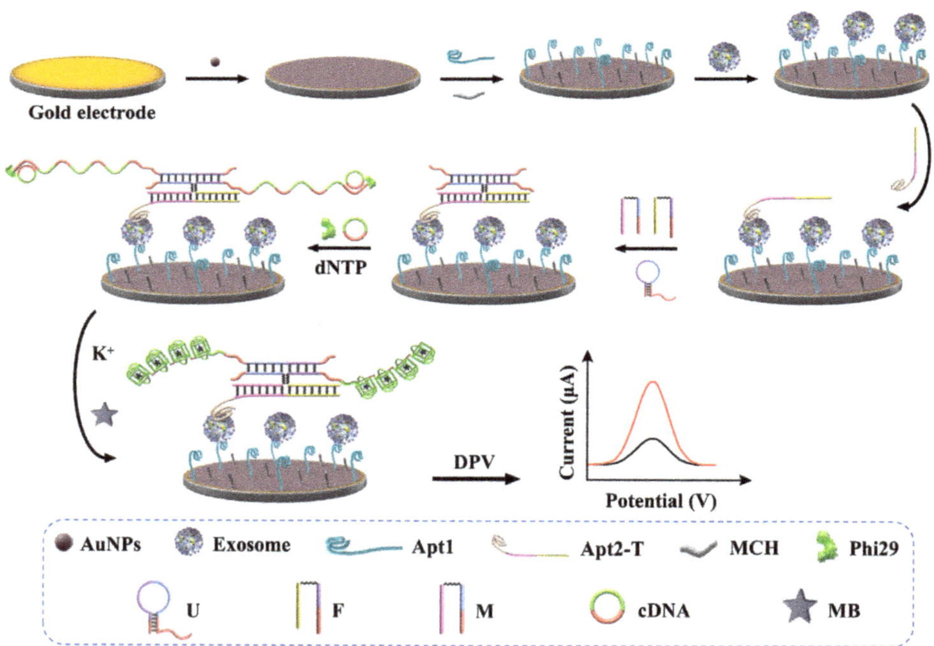

Figure 8. Schematic diagram of a bispecific aptamer sandwich-type AuNPs-modified electrochemical aptasensor for tumor-derived exosome assays based on a four-way junction-triggered dual-RCA-assisted methylene blue/G4 strategy. The repeated G4 structures produced by the RCA reaction were hanging on the electrode surface by the specific binding between split aptamer and exosome, then produced an electrochemical signal to characterize the presence of exosome (6-mercapto-1-hexanol (MCH), strand U (U), strand F (F), strand M (M), methylene blue (MB), differential pulse voltammetry (DPV)) [51].

Electrochemical aptasensors are widely employed for detecting and quantifying protein biomarkers, yet their practical use is often constrained by ineffective signal amplification and the complex, time-consuming process of probe surface immobilization. To overcome these challenges, Qing et al. integrated a homogeneous electrochemical aptasensor with the CRISPR/Cas12a system [52]. In their approach, the binding-induced DNA strand displacement (Figure 9A) mechanism is utilized to convert the interaction between thrombin and its aptamer into a nucleic acid signal, which then activates RCA to regulate the nuclease activity of CRISPR/Cas12a (Figure 9B). By exploiting the differential affinities of graphene for single-stranded versus double-stranded DNA, the electrochemical aptasensor eliminates the need for probe immobilization, simplifying the detection process and enhancing its versatility. The combination of binding-induced DNA strand displacement with RCA and CRISPR/Cas12a amplification techniques enables this aptasensor to achieve highly specific and sensitive detection of thrombin, with a limit of detection as low as 1.26 fM. This approach paves the way for more precise and efficient protein biomarker detection.

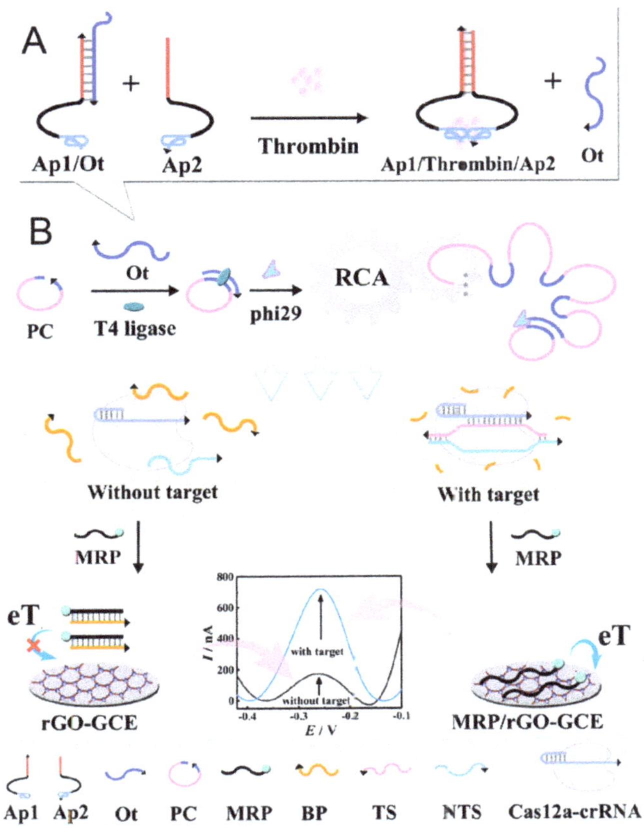

Figure 9. Schematic illustration of the working mechanism of this electrochemical aptasensor for thrombin assay: (**A**) thrombin-triggered binding-induced DNA strand displacement; (**B**) RCA regulated CRISPR/Cas12a system released MRP that can be absorbed on electrode surface to produce an electrochemical signal (output strand (Ot), padlock probe (PC), methylene blue-modified reporter probe (MRP), blocker probe (BP), trigger DNA strand (TS), non-trigger DNA strand (NTS)) [52].

An aptasensor is presented for electrochemical determination of OTA by the Abnous group based on non-target-triggered production of RCA. The surface of the gold electrode is modified with a thiolated complementary strand of aptamer as both the capture probe and primer and OTA aptamer as both the sensing molecule and padlock probe. Following the addition of OTA, aptamer/OTA conjugate is formed and detached from the electrode surface. Therefore, no RCA is produced after incubation of the modified electrode with T4 DNA ligase and phi29 DNA polymerase, and a sharp current signal occurs. The analytical response ranged from 30 pM to 120 nM with a detection limit of 5 pM [53].

As shown in Table 1, combined with the elaboration about aptamer-RCA biosensing systems, we concluded the features, advantages, and limitations of the four detection methods.

Table 1. Comparative summary of different aptamer-RCA biosensing platforms.

Detection Method	Sensitivity	Complexity	Instrument Requirement	Key Advantages	Limitations
Fluorescent	High (fM–pM)	Medium	Fluorescence reader	High sensitivity; visual readout; multiplexing	Prone to backgrou noise; high equipm cost
Electrochemical	Very high (aM–fM)	High	Potentiostat	Portable; excellent sensitivity and specificity	Surface modificati complexity; reproducibility issu
Visual	Moderate (nM–pM)	Low	None/Basic	Simple; naked-eye detection	Lower sensitivity; o qualitative/semi quantitative
Microfluidic	High (pM–fM)	High	Microfluidic chip system	Integrated, fast, miniaturized	High fabrication cc needs standardizat

6. Summary and Outlook

The highly efficient amplification characteristics of RCA make it a powerful tool for enhancing the sensitivity and specificity of biosensors, particularly in the detection of low-abundance target molecules such as viral RNAs (e.g., SARS-CoV-2 RNA), small-molecule toxins (e.g., OTA, aflatoxin B1), and cell surface markers (e.g., CD63 on exosomes). Aptamers, as highly specific synthetic nucleic acid molecules, can selectively recognize diverse targets including proteins (e.g., thrombin, mucin-1, platelet-derived growth factor), metabolites (e.g., ATP, glutathione), and whole cells or pathogens (e.g., *E. coli*, Salmonella typhimurium), thereby effectively initiating RCA for signal amplification and enabling efficient recognition and quantification of the target analytes (As shown in Table 2).

Currently, the integration of RCA with aptamers has been widely applied in the development of various sensors, particularly in areas such as food safety, environmental monitoring, and disease diagnostics. These sensors offer not only exceptional sensitivity and specificity but also the capability to perform precise detection in complex samples. However, several challenges remain. First, the selectivity of aptamers may be compromised in complex biological matrices where structurally similar molecules (e.g., serum proteins, endogenous nucleases) coexist with the intended targets. Second, RCA reactions may suffer from instability or reduced efficiency under suboptimal conditions, such as low magnesium ion concentration or the presence of polymerase inhibitors. Third, interference from non-target molecules—such as non-specific DNA-binding proteins, competing nucleic acids, or serum albumin—can affect both aptamer binding and amplification fidelity, especially in clinical or environmental samples. Addressing these limitations will require multidisciplinary efforts, including the development of chemically modified aptamers with enhanced binding affinity and nuclease resistance; optimization of RCA polymerase systems for robustness in biological fluids; and integration with orthogonal technologies such as CRISPR/Cas-based detection, nanomaterial-enhanced signal readouts, and microfluidic point-of-care devices. While the integration of aptamers and RCA has demonstrated promising diagnostic capabilities, critical challenges remain. Many current platforms rely on proof-of-concept systems, and their transition to clinical or field use is hindered by complexity, stability concerns, and a lack of standardization across platforms. In particular, trade-offs between sensitivity and operational simplicity must be carefully managed in future research.

Table 2. Aptamer–RCA integrated biosensing systems.

Aspect	Description	Examples/Notes
Principle	Combination of aptamer-based target recognition with RCA for signal enhancement	RCA generates long DNA products to amplify detection signals upon target recognition by aptamers
Key targets detected	Viral RNAs Small-molecule toxins Cell surface markers Proteins and metabolites Whole cells or pathogens	SARS-CoV-2 RNA OTA, Aflatoxin B1 CD63 on exosomes Thrombin, Mucin-1, PDGF, ATP, Glutathione *E. coli*, Salmonella typhimurium
Application areas	Disease diagnostics Food safety Environmental monitoring	Effective even in complex biological and environmental samples
Advantages	High sensitivity High specificity Signal amplification without thermal cycling Potential for miniaturization and portability	Aptamers provide target recognition; RCA ensures signal amplification
Challenges	Reduced aptamer selectivity in complex samples Inefficient RCA in suboptimal reaction conditions Interference from non-target biomolecules	Serum proteins or nucleases may bind non-specifically Low Mg^{2+} or inhibitors reduce RCA activity DNA-binding proteins or albumin disrupt performance
Strategies to Overcome Challenges	Chemically modify aptamers for better stability and binding Engineer polymerases for robust RCA in biological fluids Integrate with complementary technologies	CRISPR/Cas detection Nanomaterials (e.g., gold NPs, graphene oxide) Microfluidic or portable systems
Future Directions	Real-time biomarker monitoring Single-cell analysis On-site toxin detection using portable sensors	Emphasis on simplicity, standardization, cost-effectiveness, and clinical applicability

Looking ahead, the integration of RCA and aptamers holds great potential to revolutionize next-generation biosensors, particularly in making molecular diagnostics more rapid, accessible, and application-oriented. However, realizing this potential requires targeted advancements across several fronts. First, improving the stability and performance of aptamers in real-world samples remains a key challenge. In clinical or environmental samples, aptamers often face interference from abundant biomolecules such as serum proteins or nucleases. To address this, future efforts may focus on designing chemically modified aptamers that exhibit higher binding affinity, improved structural stability, and resistance to enzymatic degradation. Second, enhancing the robustness of RCA under physiological conditions is essential for its practical use outside of controlled laboratory settings. The development of engineered DNA polymerases with improved tolerance to inhibitors and low ionic strength, or even enzyme-free RCA alternatives using DNAzymes, could help achieve more reliable signal amplification in complex matrices like blood, urine, or food extracts. Third, significant progress can be made by integrating RCA-based biosensors with advanced signal transduction and device platforms, such as the following: Nanomaterials such as gold nanoparticles, carbon nanotubes, and graphene oxide can enhance signal output via fluorescence quenching or electrochemical amplification. Microfluidic chips can automate sample processing and reaction steps, making the entire detection process faster and more user-friendly. Portable devices, including smartphone-based readout systems, can enable point-of-care applications for infectious disease screening or food contamination detection in resource-limited settings. Moreover, RCA-aptamer platforms could be extended to real-time and multiplexed detection. For example, RCA circuits could be engineered to encode unique fluorescent barcodes, allowing simultaneous detection of multiple biomarkers from a single sample, which is a capability crucial for early disease diagnosis and personalized medicine. Lastly, with the growing interest in single-cell analysis, there is great potential to adapt aptamer–RCA strategies for detecting cell-specific surface markers or intracellular molecules at the individual cell level. This would open up new opportunities in cancer diagnostics, immunophenotyping, and stem cell research.

Funding: This research received no external funding.

Data Availability Statement: Data will be made available on request.

Conflicts of Interest: The authors declare no conflicts of interest.

Abbreviations

The following abbreviations are used in this manuscript:

functional nucleic acids	FNA
systematic evolution of ligands by exponential enrichment	SELEX
rolling circle amplification	RCA
capillary electrophoresis–SELEX	CE-SELEX
protein microarray system–SELEX	PMM-SELEX
circular DNA template	CDT
deoxyribonucleoside triphosphates	dNTPs
RNA-cleaving DNAzyme	RCD
ochratoxin A	OTA
G-quadruplex	G4
aptamer probe–hairpin primer probe	APH
time-resolved fluorescence nanoparticles	TRFNPs
silver nanoclusters	AgNCs
Salmonella typhimurium	S.T.
Vibrio parahaemolyticus	V.P.

References

1. He, W.; Liu, C.; Han, J.; Wang, S.; Lv, Y.; Li, X.; Guo, X. MNPs-aptazymes-PtNPs molecular motor biosensor mediated by ci cleavage reactions for the ultrasensitive detection of aflatoxin B1 in real samples. *Microchem. J.* **2024**, *207*, 112007. [CrossR
2. Curtis, E.A.; Liu, D.R. A naturally occurring, noncanonical GTP aptamer made of simple tandem repeats. *RNA Biol.* **20** 682–692. [CrossRef] [PubMed]
3. Wu, Y.X.; Kwon, Y.J. Aptamers: The "evolution" of SELEX. *Methods* **2016**, *106*, 21–28. [CrossRef]
4. Singer, B.S.; Shtatland, T.; Brown, D.; Gold, L. Libraries for genomic SELEX. *Nucleic Acids Res.* **1997**, *25*, 781–786. [CrossRe
5. Hu, Y.-Y.; Yang, G.; Qu, F. Research advances in non-immobilized aptamer screening techniques for small-molecule targets *J. Chromatogr.* **2025**, *43*, 297–308. [CrossRef]
6. Ellington, A.D.; Szostak, J.W. In vitro selection of RNA molecules that bind specific ligands. *Nature* **1990**, *346*, 818–822. [Cro
7. Tsuji, S.; Hirabayashi, N.; Kato, S.; Akitomi, J.; Egashira, H.; Tanaka, T.; Waga, I.; Ohtsu, T. Effective isolation of RNA ap through suppression of PCR bias. *Biochem. Biophys. Res. Commun.* **2009**, *386*, 223–226. [CrossRef]
8. Yang, X.; Chan, C.H.; Yao, S.; Chu, H.Y.; Lyu, M.; Chen, Z.; Xiao, H.; Ma, Y.; Yu, S.; Li, F.; et al. DeepAptamer: Adva high-affinity aptamer discovery with a hybrid deep learning model. *Mol. Ther. Nucleic Acids* **2025**, *36*, 102436. [CrossRef]
9. Wang, W.; Jia, L.Y. Progress in aptamer screening methods. *Chin. J. Anal. Chem.* **2009**, *37*, 454–460. [CrossRef]
10. Hatch, A.; Sano, T.; Misasi, J.; Smith, C.L. Rolling circle amplification of DNA immobilized on solid surfaces and its applica multiplex mutation detection. *Genet. Anal. Biomol. Eng.* **1999**, *15*, 35–40. [CrossRef]

Boonbanjong, P.; Treeratrakoon, K.; Waiwinya, W.; Pitikultham, P.; Japrung, D. Isothermal amplification technology for disease diagnosis. *Biosensors* **2022**, *12*, 677. [CrossRef] [PubMed]
Ou, L.; Sun, A.; Liu, K. Rolling circle amplification-based biosensors. *Anal. Lett.* **2015**, *48*, 1199–1216. [CrossRef]
Zhan, Y.; Mao, Y.; Sun, P.; Liu, C.; Gou, H.; Qi, H.; Chen, G.; Hu, S.; Tian, B. Tumor-associated antigen-specific cell imaging based on upconversion luminescence and nucleic acid rolling circle amplification. *Microchim. Acta* **2024**, *191*, 248. [CrossRef] [PubMed]
Zhang, X.-W.; Qi, G.-X.; Chen, S.; Yu, Y.-L.; Wang, J.-H. Ultrasensitive and wash-free detection of tumor extracellular vesicles by aptamer-proximity-ligation-activated rolling circle amplification coupled to single particle ICP-MS. *Anal. Chem.* **2024**, *96*, 10800–10808. [CrossRef]
Tang, J.; Yu, Y.; Shi, H.; He, X.; Lei, Y.; Shangguan, J.; Yang, X.; Qiao, Z.; Wang, K. Polyvalent and thermosensitive DNA nanoensembles for cancer cell detection and manipulation. *Anal. Chem.* **2017**, *89*, 6637–6644. [CrossRef]
Zhu, D.; Huang, T.; Zhou, Q.; Yang, Z.; Liu, B.; Li, M.; Li, C.; Chen, J.-X.; Dai, Z.; Chen, J. A label-free fluorescent aptasensor based on a novel exponential rolling circle amplification for highly sensitive ochratoxin A detection. *Food Chem.* **2023**, *410*, 135427. [CrossRef]
Xing, Y.; Wang, P.; Zang, Y.; Ge, Y.; Jin, Q.; Zhao, J.; Xu, X.; Zhao, G.; Mao, H. A colorimetric method for H1N1 DNA detection using rolling circle amplification. *Analyst* **2013**, *138*, 3457–3462. [CrossRef]
Ding, C.; Liu, H.; Wang, N.; Wang, Z. Cascade signal amplification strategy for the detection of cancer cells by rolling circle amplification and nanoparticles tagging. *Chem. Commun.* **2012**, *48*, 5019–5021. [CrossRef]
Zhu, Y.; Wang, H.; Wang, L.; Zhu, J.; Jiang, W. Cascade signal amplification based on copper nanoparticle-reported rolling circle amplification for ultrasensitive electrochemical detection of the prostate cancer biomarker. *ACS Appl. Mater. Interfaces* **2016**, *8*, 2573–2581. [CrossRef]
Kumar, Y. Isothermal amplification-based methods for assessment of microbiological safety and authenticity of meat and meat products. *Food Control* **2021**, *121*, 107679. [CrossRef]
Lau, H.Y.; Botella, J.R. Advanced DNA-based point-of-care diagnostic methods for plant diseases detection. *Front. Plant Sci.* **2017**, *8*, 2016. [CrossRef] [PubMed]
Tomas Rozenblum, G.; Gisela Lopez, V.; Daniel Vitullo, A.; Radrizzani, M. Aptamers: Current challenges and future prospects. *Expert Opin. Drug Discov.* **2016**, *11*, 127–135. [CrossRef] [PubMed]
Ali, M.H.; Elsherbiny, M.E.; Emara, M. Updates on aptamer research. *Int. J. Mol. Sci.* **2019**, *20*, 2511. [CrossRef]
Du, Y.-L.; Mo, L.-T.; Yi, Y.-S.; Qiu, L.-P.; Tan, W.-H. Aptamers from cell-based selection for bioanalysis and bioimaging. *Chin. J. Anal. Chem.* **2017**, *45*, 1757–1764. [CrossRef]
Yang, G.; Han, S.-M.; Zhao, L.-P.; Zhu, C.; Huang, Y.-Y.; Qu, F. Screening aptamer of apo-transferrin via capillary electrophoresis-systematic evolution of ligands by exponential enrichment and environmental factors analysis. *Chin. J. Anal. Chem.* **2020**, *48*, 632–641. [CrossRef]
Liu, X.; Li, H.; Jia, W.; Chen, Z.; Xu, D. Selection of aptamers based on a protein microarray integrated with a microfluidic chip. *Lab. Chip* **2017**, *17*, 178–185. [CrossRef]
Chen, J.; Zhou, J.; Peng, Y.; Xie, Y.; Xiao, Y. Aptamers: A prospective tool for infectious diseases diagnosis. *J. Clin. Lab. Anal.* **2022**, *36*, 24725. [CrossRef]
Rothlisberger, P.; Gasse, C.; Hollenstein, M. Nucleic acid aptamers: Emerging applications in medical imaging, nanotechnology, neurosciences, and drug delivery. *Int. J. Mol. Sci.* **2017**, *18*, 2430. [CrossRef]
Catherine, A.T.; Shishido, S.N.; Robbins-Welty, G.A.; Diegelman-Parente, A. Rational design of a structure-switching DNA aptamer for potassium ions. *FEBS Open Bio* **2014**, *4*, 788–795. [CrossRef]
Bialy, R.M.; Mainguy, A.; Li, Y.; Brennan, J.D. Functional nucleic acid biosensors utilizing rolling circle amplification. *Chem. Soc. Rev.* **2022**, *51*, 9009–9067. [CrossRef]
Chai, S.; Sun, W.; Hou, X.; Pei, S.; Liu, Y.; Luo, K.; Guan, S.; Lv, W. A primer-regulated rolling circle amplification (RCA) for logic-controlled multiplexed enzyme analysis. *Acs Appl. Bio Mater.* **2025**, *8*, 2408–2418. [CrossRef] [PubMed]
Liu, M.; Song, J.; Shuang, S.; Dong, C.; Brennan, J.D.; Li, Y. A graphene-based biosensing platform based on the release of DNA probes and rolling circle amplification. *ACS Nano* **2014**, *8*, 5564–5573. [CrossRef] [PubMed]
Liu, M.; Zhang, Q.; Chang, D.; Gu, J.; Brennan, J.D.; Li, Y. A DNAzyme feedback amplification strategy for biosensing. *Angew. Chem. Int. Ed.* **2017**, *56*, 6142–6146. [CrossRef] [PubMed]
Sakhabutdinova, A.R.; Maksimova, M.A.; Garafutdinov, R.R. Synthesis of circular DNA templates with T4 RNA ligase for rolling circle amplification. *Mol. Biol.* **2017**, *51*, 724–733. [CrossRef]
Di Giusto, D.A.; Wlassoff, W.A.; Gooding, J.J.; Messerle, B.A.; King, G.C. Proximity extension of circular DNA aptamers with real-time protein detection. *Nucleic Acids Res.* **2005**, *33*, e64. [CrossRef]
Wang, L.; Tram, K.; Ali, M.M.; Salena, B.J.; Li, J.; Li, Y. Arrest of rolling circle amplification by protein- binding DNA aptamers. *Chem. Eur. J.* **2014**, *20*, 2420–2424. [CrossRef]

37. Bialy, R.M.; Ali, M.M.; Li, Y.; Brennan, J.D. Protein-mediated suppression of rolling circle amplification for biosensing wi aptamer-containing DNA primer. *Chem. Eur. J.* **2020**, *26*, 5085–5092. [CrossRef]
38. Song, J.; Ju, Y.; Kim, S.; Kim, H.; Park, H.G. Palindromic hyperbranched rolling circle amplification enabling ultrasens microRNA detection. *Chem. Commun.* **2022**, *58*, 6518–6521. [CrossRef]
39. Lu, P.; Aodeng, G.; Ai, J. Progress in ion detection and clinical diagnosis based on G-quadruplex combined with fluoresc properties. *Adv. Sens. Energy Mater.* **2024**, *3*, 100112. [CrossRef]
40. Wang, W.; Geng, L.; Zhang, Y.; Shen, W.; Bi, M.; Gong, T.; Hu, Z.; Guo, C.; Wang, T.; Sun, T. Development of antibody-apt sandwich-like immunosensor based on RCA and Nicked-PAM CRISPR/Cas12a system for the ultra-sensitive detection biomarker. *Anal. Chim. Acta* **2023**, *1283*, 341849. [CrossRef]
41. Niazi, S.; Khan, I.M.; Yu, Y.; Pasha, I.; Lv, Y.; Mohsin, A.; Mushtaq, B.S.; Wang, Z. A novel fluorescent aptasensor for afla M_1 detection using rolling circle amplification and g-C_3N_4 as fluorescence quencher. *Sens. Actuators B Chem.* **2020**, *315*, 12 [CrossRef]
42. Zhang, M.; Ye, L. Detection of SARS-CoV-2 receptor binding domain using fluorescence probe and DNA flowers enable rolling circle amplification. *Microchim. Acta* **2023**, *190*, 163. [CrossRef] [PubMed]
43. Zeng, Y.; Qi, P.; Zhou, Y.; Wang, Y.; Xin, Y.; Sun, Y.; Zhang, D. Multi pathogenic microorganisms determination using I composites-encapsulated DNA silver nanocluster/graphene oxide-based system through rolling cycle amplification. *Micro Acta* **2022**, *189*, 403. [CrossRef]
44. Jiang, Y.; Qiu, Z.; Le, T.; Zou, S.; Cao, X. Developing a dual-RCA microfluidic platform for sensitive *E. coli O* 157:H7 whol detections. *Anal. Chim. Acta* **2020**, *1127*, 79–88. [CrossRef]
45. Li, S.; Jiang, Y.; Yang, X.; Lin, M.; Dan, H.; Zou, S.; Cao, X. In situ rolling circle amplification surface modifications to impro *coli* O157:H7 capturing performances for rapid and sensitive microfluidic detection applications. *Anal. Chim. Acta* **2021**, 338229. [CrossRef]
46. Yu, J.; Wu, H.; He, L.; Tan, L.; Jia, Z.; Gan, N. The universal dual-mode aptasensor for simultaneous determination of diff bacteria based on naked eyes and microfluidic-chip together with magnetic DNA encoded probes. *Talanta* **2021**, *225*, 12 [CrossRef]
47. Gao, X.; Sun, Z.; Wang, X.; Zhang, W.; Xu, D.; Sun, X.; Guo, Y.; Xu, S.; Li, F. Construction of a dual-model aptasensor base G-quadruplexes generated via rolling circle amplification for visual/sensitive detection of kanamycin. *Sci. Total Environ.* *839*, 156276. [CrossRef]
48. Wang, X.-Y.; Yan, Y.; Zhi, S.; Bi, S. Dumbbell DNA-mediated rolling circle amplification for visual biosensing of intrace glutathione. *Sens. Actuators B Chem.* **2022**, *373*, 132745. [CrossRef]
49. Zhang, H.; Mao, W.; Hu, Y.; Wei, X.; Huang, L.; Fan, S.; Huang, M.; Song, Y.; Yu, Y.; Fu, F. Visual detection of aflatoxin B1 base specific aptamer recognition combining with triple amplification strategy. *Spectrochim. Acta Part A Mol. Biomol. Spectrosc.* *271*, 120862. [CrossRef]
50. Yurdusev, E.; Trahan, P.-L.; Perreault, J. Adaptation of a model spike aptamer for isothermal amplification-based sensing. *Se* **2024**, *24*, 6875. [CrossRef]
51. Zhao, Z.; Yang, S.; Tang, X.; Feng, L.; Ding, Z.; Chen, Z.; Luo, X.; Deng, R.; Sheng, J.; Xie, S.; et al. DNA four-way junction-d dual-rolling circle amplification sandwich-type aptasensor for ultra-sensitive and specific detection of tumor-derived exos *Biosens. Bioelectron.* **2024**, *246*, 115841. [CrossRef] [PubMed]
52. Qing, M.; Sun, Z.; Wang, L.; Du, S.Z.; Zhou, J.; Tang, Q.; Luo, H.Q.; Li, N.B. CRISPR/Cas12a-regulated homogeneous el chemical aptasensor for amplified detection of protein. *Sens. Actuators B Chem.* **2021**, *348*, 130713. [CrossRef]
53. Taghdisi, S.M.; Danesh, N.M.; Ramezani, M.; Alibolandi, M.; Nameghi, M.A.; Gerayelou, G.; Abnous, K. A novel electroche aptasensor for ochratoxin a sensing in spiked food using strand-displacement polymerase reaction. *Talanta* **2021**, *223*, 12 [CrossRef] [PubMed]

molecules

w

nctional Nucleic Acid Nanostructures for Mitochondrial rgeting: The Basis of Customized Treatment Strategies

chcng He *, Siyu Dong and Qinghua Zeng *

School of Pharmaceutical Sciences and Food Engineering, Liaocheng University, Liaocheng 252059, China
* Correspondence: hewanchong@gmail.com (W.H.); zengqinghua@lcu.edu.cn (Q.Z.)

Abstract: Mitochondria, as vital organelles, play a central role in subcellular research and biomedical innovation. Although functional nucleic acid (FNA) nanostructures have witnessed remarkable progress across numerous biological applications, strategies specifically tailored to target mitochondria for molecular imaging and therapeutic interventions remain scarce. This review delves into the latest advancements in leveraging FNA nanostructures for mitochondria-specific imaging and cancer therapy. Initially, we explore the creation of FNA-based biosensors localized to mitochondria, enabling the real-time detection and visualization of critical molecules essential for mitochondrial function. Subsequently, we examine developments in FNA nanostructures aimed at mitochondrial-targeted cancer treatments, including modular FNA nanodevices for the precise delivery of therapeutic agents and programmable FNA nanostructures for disrupting mitochondrial processes. Emphasis is placed on elucidating the chemical principles underlying the design of mitochondrial-specific FNA nanotechnology for diverse biomedical uses. Lastly, we address the unresolved challenges and outline prospective directions, with the goal of advancing the field and encouraging the creation of sophisticated FNA tools for both academic inquiry and clinical applications centered on mitochondria.

Keywords: FNA nanostructures; mitochondria; customized treatment strategy

1. Introduction

Mitochondria, often referred to as the "powerhouses" of eukaryotic cells, are responsible for generating adenosine 5′-triphosphate (ATP) via oxidative phosphorylation [1]. Beyond their fundamental role in energy production, mitochondria are integral to cellular homeostasis, contributing to metabolic regulation and responding to both internal and external stressors such as nutrient deficiencies and redox imbalances [1,2]. These organelles also manage the synthesis and removal of numerous metabolic intermediates and waste products [2]. Moreover, mitochondria act as critical regulators of programmed cell death, including apoptosis and necroptosis [3]. For instance, they store pro-apoptotic proteins and control their release into the cytoplasm to trigger apoptotic pathways [4,5]. Dysregulation of mitochondrial functions has been strongly associated with the onset and progression of various diseases, including diabetes, cancer, and neurodegenerative disorders [6]. Consequently, significant efforts are focused on the in situ analysis of specific mitochondrial components to uncover the molecular mechanisms underpinning these processes. Additionally, researchers are translating insights into mitochondrial biology into therapeutic strategies, as growing evidence highlights the superior efficacy of mitochondria-targeted treatments compared to conventional approaches [7,8].

Functional nucleic acids (FNAs), including DNA and RNA aptamers, Triplex DNA, DNAzymes, RNAzymes, DNA tiles, and DNA origami, have seen widespread use as molecular recognition elements for specific binding with various targets [9–12]. Unlike other substances, FNA offers several distinctive attributes that make it highly compatible with biological systems. Firstly, as an inherent component of life, DNA exhibits excellent biocompatibility. Secondly, FNA can be readily functionalized with diverse chemical groups, such as fluorophores, phosphorothioates, or therapeutic agents, through chemical synthesis or enzymatic modification [13]. Furthermore, the specific biorecognition capabilities of aptamers and DNAzymes have been discovered and characterized [14–16]. Notably, the Watson–Crick base pairing mechanism provides FNA with predictable thermodynamic behavior and exceptional programmability, facilitating the design of intricately shaped nanostructures and adaptive nanomachines capable of dynamic reconfiguration in response to biological stimuli [17–19]. Despite progress in FNA-based nanobiotechnology for diagnostics and imaging, mitochondria-targeted FNA systems remain underdeveloped [20–34]. This limitation primarily stems from the inability of most FNA-based nanostructures to achieve precise spatial and temporal localization within organelles, which is critical for executing specialized functions.

In recent years, researchers have been actively exploring the fusion of DNA-based technologies with organelle-specific nanotechnology to create sophisticated DNA nanodevices capable of precisely targeting mitochondria and performing their designated functions at the subcellular level [35]. Compared with other mitochondria targeting strategies (e.g., peptide-based targeting or lipid nanoparticles (LNPs), FNA nanostructures are highly programmable, biocompatible, and specific, making them ideal for applications requiring precision targeting, such as biosensing and gene editing. Meanwhile, the delivery efficiency also needs to improve by more targeted modification or by combination with the LNP technique [36,37]. This review provides an overview of the latest breakthroughs in DNA-driven nanobiotechnologies aimed at mitochondrial imaging and therapy. The goal of this review is to comprehensively summarize the current mitochondria-targeted strategies based on functional nucleic acid nanostructures, clarify their advantages over other current mitochondrial targeting methods, and explore the current shortcomings and challenges of this method as well as the direction of future improvement and development. Initially, we delve into mitochondria-specific DNA nanosensors designed for selective imaging of essential mitochondrial molecules, such as nucleic acids, enzymes, small molecules, and metal ions. Particular attention is given to the underlying design strategies of these nanosensors, which offer promising tools for investigating the intricate molecular dynamics associated with mitochondrial roles in health and disease. Next, we outline recent advancements in two key DNA-based therapeutic approaches targeting mitochondria. The first involves the use of modularly engineered DNA nanodevices to facilitate the controlled delivery of therapeutic agents into mitochondria, while the second employs in situ programmed DNA self-assembly to directly modulate mitochondrial function. Finally, we examine the existing hurdles and future prospects within this field. By consolidating the latest advancements in this promising area, we aim to inspire the development of innovative methodologies that can benefit both fundamental research and clinical applications involving mitochondria.

2. Mitochondria Targeting Strategies Based on FNA Nanostructures

2.1. FNA Targeting Mitochondrial RNA (mtRNA)

In eukaryotic cells, while the majority of DNA is tightly packed into chromatin structures and housed within the nucleus, mitochondria possess a unique subset of DNA, known as mtDNA, which distinguishes them from other organelles [38]. Beyond mtDNA, an increasing body of evidence suggests that microRNAs (miRNAs), primarily transcribed from the nuclear genome and typically active in the cytoplasm, can also localize within mitochondria, where they are referred to as mitomiRs [39,40]. These nucleic acids localized in mitochondria play critical roles in regulating mitochondrial functions. For instance, mutations in mtDNA are known to disrupt mitochondrial functionality, potentially causing various metabolic disorders [41]. Furthermore, abnormal expression levels of mitomiRs have been associated with a wide array of diseases, including neurodegenerative conditions, cardiovascular disorders, and cancers [42,43]. Some mitomiRs even hold promise as diagnostic biomarkers or therapeutic targets for these diseases [43]. Despite significant advancements in DNA-based sensing technologies for detecting and imaging nucleic acids in live cells [44–47], their application in tracking mitochondrial targets presents substantial challenges. These difficulties arise from the lack of mitochondria-specific targeting capabilities. Moreover, many nucleic acid targets, such as miRNAs, are not confined to a specific organelle. Conventional sensing approaches, which rely on an "always active" detection mechanism, risk premature activation upon encountering these targets before reaching the intended organelle, thereby compromising spatial accuracy. FNA nanostructures provide a promising approach for targeting mtDNA mutations through precise sequence recognition, targeted delivery, and controlled gene modulation. Their interaction with mtDNA mutations occurs via several mechanisms: 1. Targeted gene editing and repair. FNA nanostructures can deliver gene editing systems to mitochondria to selectively cleave mutant mtDNA, promoting the replication of wild-type mtDNA. Unlike nuclear DNA editing, mtDNA lacks efficient DNA repair mechanisms, making targeted cleavage a viable approach for shifting heteroplasmy levels. 2. Selective silencing of mutated mtDNA. FNA nanostructures designed as small interfering RNAs or mitochondria-targeted aptamers can selectively bind mutated mt-mRNAs, reducing the expression of defective proteins. Aptamer-functionalized FNAs provide high specificity in targeting mutant transcripts while sparing wild-type sequences. 3. Mitochondrial delivery of therapeutic genes. FNA nanostructures can be engineered as nanocarriers to deliver therapeutic RNA/DNA molecules, compensating for mtDNA mutations.

To address the above-mentioned issues, some FNA nanostructures which target mitochondria were developed. The Qian group introduced a novel "RT-qPCR mimic" system, incorporating a signal amplification mechanism without the need for enzymes [48]. This system employs a hairpin DNA cascade amplifier (HDCA), which is composed of two metastable hairpin DNA structures and a hybrid DNA duplex reporter (Figure 1), to independently visualize mtRNA and cytosolic reference mRNA. To enable the precise delivery of probes to subcellular targets such as mitochondria and the cytosol, black phosphorus nanosheets (BPNSs) were utilized These nanosheets, known for their excellent biocompatibility, high molecular loading efficiency [49,50], and ability to degrade into non-toxic phosphate and phosphonate ions under physiological conditions [50–54], served as stable transport platforms. Additionally, the BPNS surface's unique combination of a high surface-area-to-volume ratio and periodic atomic grooves offered optimal attachment sites for nucleic acids [55–57], shielding them from degradation by endogenous nucleases. Demonstrating the approach, the HDCA designed to detect mtRNA encoding NADH dehydrogenase subunit 6 (mtRNAND6) was loaded onto mitochondria-specific BPNS carriers, BP-PEI-TPP, for targeted delivery and detection. Simultaneously, the levels of β-actin mRNA (referred to as mRNAβ-actin) in the cytoplasm were measured as an internal control. This innovation represents the first dual-color imaging platform capable of accurately quantifying specific mitochondrial RNAs in living cells.

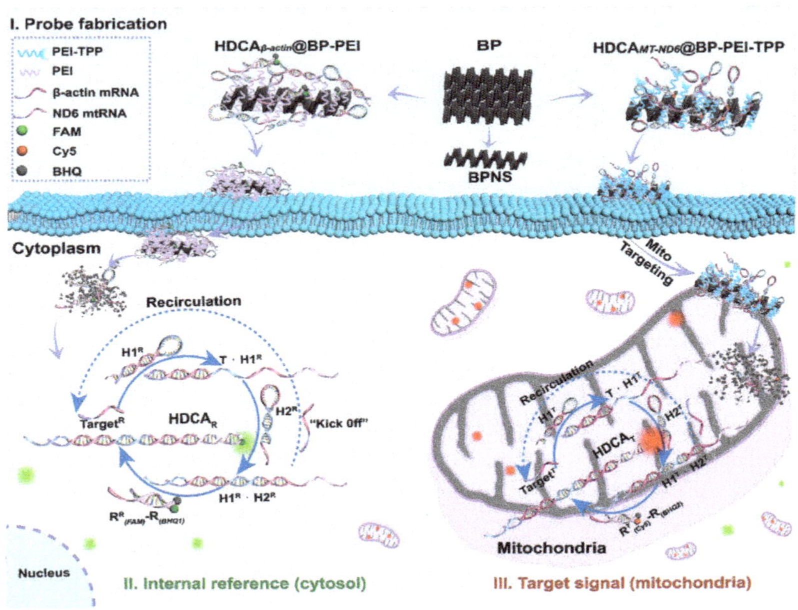

Figure 1. Schematic of the dual-color imaging system for quantitative analysis of specific mtRNA in living cells: (**I**) Fabrication of the HDCA probe onto BPNS-based vehicles for effective delivery to both the cytosol and mitochondria. (**II**) Internal reference module for detecting housekeeping gene (TargetR) in the cytoplasm. (**III**) Reporting module designed to specifically target mitochondrial RNAs (TargetT) [48].

The method employing precise spatial and temporal control serves as a versatile tool for investigating the biological roles of mitomiRs. Building on this targeting strategy, Chen et al. incorporated a fluorescence-encoded error correction system to create a near-infrared (NIR)-responsive DNA platform capable of simultaneously visualizing three mitomiRs associated with the mitochondrial mt-ND1 genome in drug-resistant cells [58]. Furthermore, a nanoscale DNA computing device was engineered to track two mitomiRs during the process of cell apoptosis [59]. More recently, the development of an AIE-labeled DNA probe, integrated with a polymer-based nanocarrier for targeted delivery, enabled the pH-sensitive exonuclease-driven imaging of mitomiRs [60].

In contrast to miRNAs, mtDNA is confined to the mitochondria and exhibits a relatively stable concentration. However, due to the elevated levels of reactive oxygen species (ROS) within mitochondria and the absence of histone protection, mtDNA is particularly vulnerable to oxidative damage, which can result in pathogenic mutations. Importantly, mtDNA mutations are highly dynamic and dispersed across the entire genome, making the detection of mutations present in low abundance at specific loci particularly challenging. To address this, Zhang et al. recently developed an integrated nanoscale Cas12a sensor (referred to as InCasor). This system employs a DNA/Mg^{2+} hybrid nanoflower (DNF) functionalized with aptamers to deliver Cas12a-related components directly to mitochondria. These components include the Cas12a/crRNA complex and a circular reporter (CLR) labeled with a fluorophore and quencher, enabling the identification of mtDNA mutations in live cells and in vivo (Figure 2) [61].

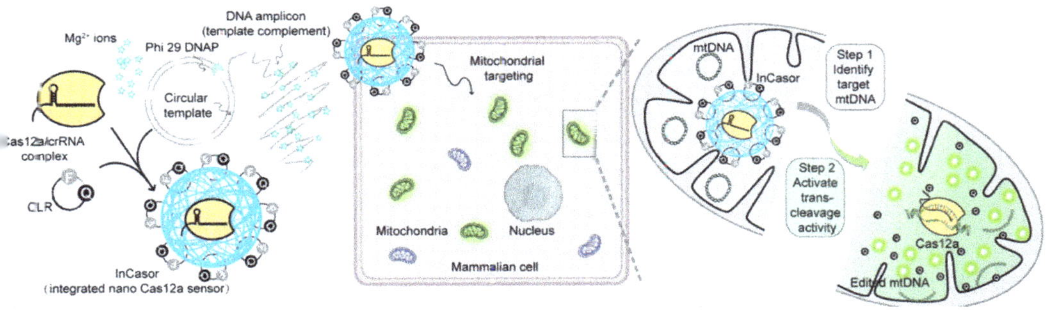

Figure 2. Schematic of the InCasor: This system comprises a DNF, Cas12a/crRNA complex, and a CLR for imaging mtDNA in live cells [61]. Through the strategic design of crRNA, single nucleotide variants in mitochondrial DNA can be precisely recognized. Additionally, increasing the intracellular concentration of Mg^{2+} significantly boosts the collateral trans-cleavage activity of Cas12a, thereby amplifying the detection signal within living cells.

A significant achievement of this research is the successful mitochondria-specific delivery of the Cas12a/crRNA complex using InCasor. The mitochondrial bilayer membrane poses a significant barrier, making the transport of guide RNA and Cas proteins into mitochondria notoriously difficult [62]. To tackle this obstacle, this study employed a nanoparticle-based delivery approach, a strategy that has proven effective in transporting proteins and nucleic acids into mitochondria for a variety of diagnostic and therapeutic purposes [63,64]. This method often leverages mitochondrial targeting moieties, such as triphenylphosphine [65]. In this investigation, nucleic acid-based nanocarriers equipped with Cyt C aptamers were developed to facilitate the targeted delivery of Cas12a/crRNA complexes to mitochondria. The experimental results demonstrated that InCasor successfully delivered these complexes into mitochondria by modulating the permeability of the mitochondrial membrane.

InCasor represents a breakthrough technology for the direct visualization of cells harboring mtDNA mutations in vivo, addressing a longstanding challenge. The ability to monitor mtDNA mutations provides critical insights into mtDNA heterogeneity and facilitates the detection of tumor tissues carrying specific mtDNA alterations. Future advancements will aim to expand InCasor's applications to include nuclear genome analysis and the precise tracking of single nucleotide variant (SNV) sites with high spatiotemporal resolution. These efforts will further establish InCasor as a versatile tool for fundamental research on gene mutations, as well as for diagnostic and therapeutic applications involving in vivo gene editing.

By incorporating patient-specific antisense DNA sequences into an FNA nanostructure to selectively bind mutant mtRNA transcripts, this strategy could minimize off-target effects and ensures that the treatment is only activated in dysfunctional mitochondria. By designing an FNA nanostructure tailored to the patient's unique mtDNA mutation profile, this approach provides high specificity, minimal off-target effects, and an adaptive therapeutic response. Over time, a shift in the heteroplasmy ratio towards wild-type mtDNA is expected, improving mitochondrial function and alleviating disease symptoms.

2.2. FNA Targeting Specific Enzymes in Mitochondria

Proteomic analyses reveal that mitochondria harbor approximately 1000 to 1500 distinct proteins, with only 13 encoded by mtDNA. The remaining proteins are encoded in nuclear DNA, synthesized in the cytoplasm, and subsequently imported into mitochondria [66]. This reliance underscores the importance of dynamic protein translocation across subcellular compartments for mitochondrial functionality. For example, various nucleases, crucial for mtDNA repair and mitochondrial RNA metabolism, are distributed between mitochondria and other organelles [67]. Human apurinic/apyrimidinic endonuclease 1 (APE1), a versatile enzyme involved in DNA repair, redox signaling, and transcription factor regulation [68,69], predominantly resides in the nucleus but can relocate to other compartments, such as mitochondria, under pathological conditions [70]. Developing a precise toolkit to monitor the intracellular dynamics of APE1 is therefore critical for advancing our understanding of disease mechanisms [71,72]. However, designing such a toolkit with the required subcellular accuracy remains a significant challenge.

Inspired by this, Li's group introduced a highly modular nanoplatform, UR-HAPT, capable of imaging subcellular APE1 dynamics in response to mitochondria-localized photodynamic therapy (PDT) activated by near-infrared (NIR) light (Figure 3) [73]. UR-HAPT is composed of four key elements: (1) an upconversion nanoparticle (UCNP) serving as a light transducer; (2) rose bengal (RB) functioning as a photosensitizer (PS); (3) triphenylphosphonium (TPP) acting as a mitochondria-targeting ligand; and (4) a DNA-based fluorescence reporter (HAP) designed for detecting APE1 enzymatic activity. The HAP probe incorporates an apurinic/apyrimidinic (AP) site within the stem of a molecular beacon, which is dual-labeled with the fluorophore Cy5 at the 3′-end and the quencher BHQ-2 at the 5′-end. This configuration ensures an exceptionally low fluorescence background through Förster resonance energy transfer (FRET) between Cy5 and BHQ-2. In the presence of APE1, cleavage at the AP site releases the quencher, resulting in a substantial fluorescence signal recovery. The TPP ligands on UR-HAPT facilitate precise mitochondrial targeting, while NIR light irradiation triggers the production of subcellular reactive oxygen species (ROS) via energy transfer from UCNPs to the PSs. Crucially, the HAP probe allows real-time tracking of APE1 translocation dynamics within mitochondria during the PDT process. This innovative design bridges the gap between photodynamic therapy and DNA-based biosensing technologies, enabling real-time subcellular molecular imaging during therapeutic interventions.

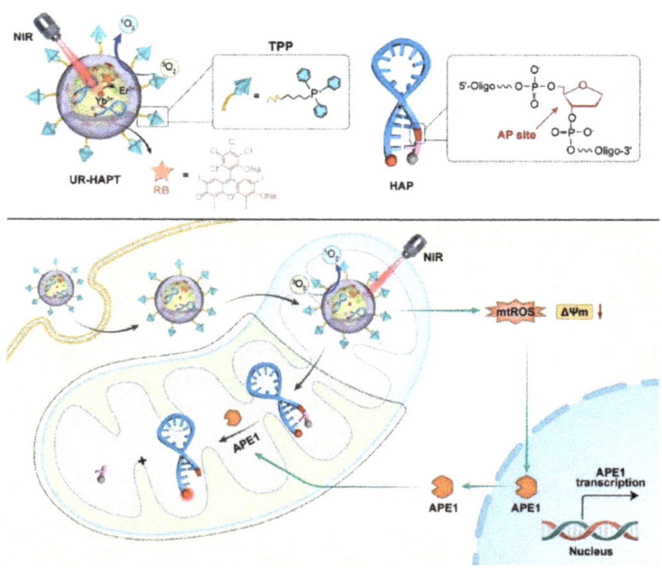

Figure 3. Schematic of the FNA nanosensor for in situ localization and NIR-light-activatable imaging of APE1 in mitochondria [73].

2.3. FNA Targeting Small Molecules and Metal Ions in Mitochondria

Mitochondria, often referred to as cellular powerhouses, generate ATP via oxidative phosphorylation [74]. However, during ATP production in cancer cells, abnormal accumulation of mitochondrial reactive oxygen species (ROS) has been observed [75,76]. Simultaneously, cancer cells synthesize large amounts of glutathione (GSH) within mitochondria to counteract oxidative stress caused by excessive ROS, thereby facilitating tumor progression [77,78]. Consequently, the disruption of energy metabolism and redox imbalance in mitochondria are widely recognized as defining characteristics of tumor development [79]. Investigating ATP and GSH—two critical molecules linked to mitochondrial energy metabolism and redox balance—through correlated imaging could shed light on their roles in tumorigenesis. While various DNA-based methods have been designed for the individual detection of ATP and GSH within cells [80,81], the simultaneous imaging of these two pivotal molecules remains an uncharted area of research.

Recently, Li's group developed a novel approach that integrates a redox-sensitive aptamer sensor with nanoparticles engineered for precise targeting, enabling spatially resolved, AND-gated visualization of ATP and GSH within mitochondria [82]. As shown in Figure 4A, the A-G/NT system consists of two interconnected components: a redox-responsive aptamer probe (A-G) for dual molecular sensing and an organelle-specific nanoparticle (NP) for targeted mitochondrial delivery. The A-G probe was assembled by hybridizing two carefully designed DNA strands: the ATP-binding A-strand, derived from an aptamer, and the G-strand, which incorporates a strategically placed disulfide bond within its DNA backbone. This hybridization positioned Cy3 (attached to the A-strand) in close proximity to a black hole quencher (anchored to the G-strand), producing a low fluorescence background due to Förster resonance energy transfer (FRET). When GSH cleaves the disulfide bond in the G-strand, its binding affinity to the A-strand diminishes, reinstating the aptamer's structure-switching ability to interact with ATP. Upon ATP recognition, the formation of an aptamer–ATP complex causes the cleaved G-strand to detach from the A-strand, leading to a fluorescence signal increase. Consequently, this stepwise activation process necessitates the concurrent presence of GSH and ATP to enable signal output, implementing an AND-gated mechanism for molecular imaging. Additionally, the A-G/NT design incorporates triphenylphosphonium (TPP) conjugation on the NPs, facilitating targeted delivery of the A-G probe for precise imaging within mitochondria [82].

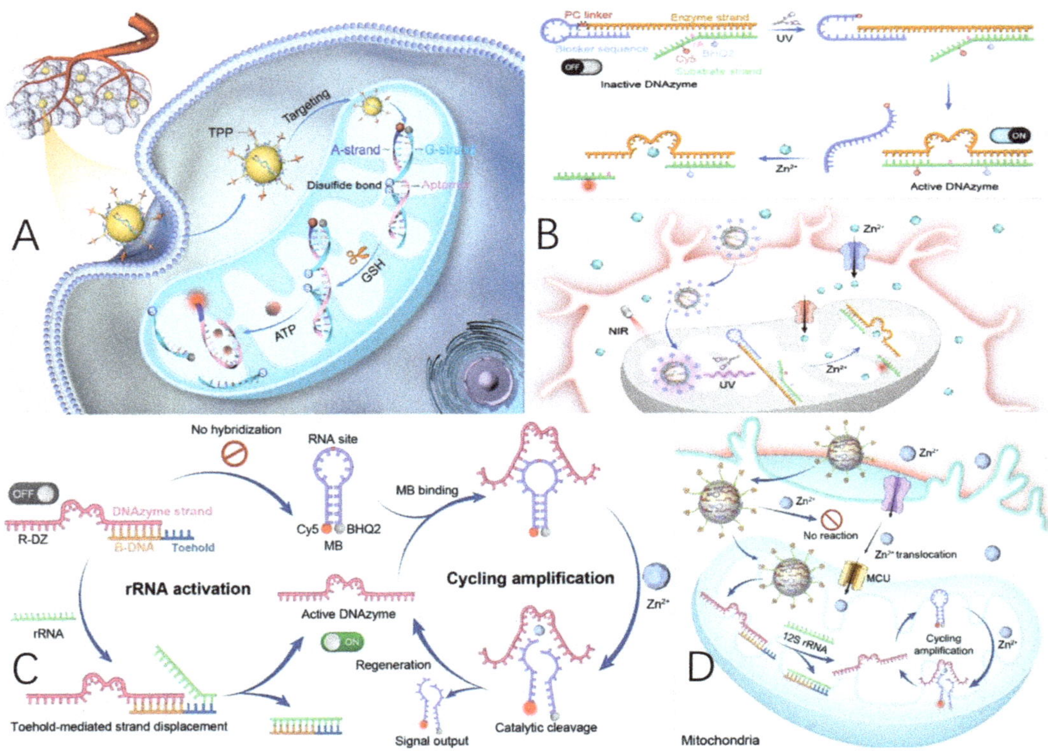

Figure 4. (**A**) Schematic of the design of a redox-activatable DNA nanodevice for AND-gated imaging of ATP and GSH in mitochondria [82]. (**B**) Schematic of the design and metal-ion sensing mechanism of L-DZ and the application of L-DZ/mUC for spatially selective imaging of $[Zn^{2+}]_m$ in Zn^{2+}-induced neuropathology [83]. (**C**) The working principle of R-DZ/MB [84]. (**D**) The rRNA-regulated DNAzyme nanosensors for subcellular compartment-specific amplified imaging of Zn^{2+} in mitochondria [84].

Metal ions are indispensable for nearly all biological activities, with their precise distribution within cellular compartments being tightly controlled [85]. Investigating the localization and behavior of these ions at the subcellular level is crucial for understanding their roles in both normal physiological and disease-related mechanisms. For instance, mitochondria have been recognized as key reservoirs of Zn^{2+}, and maintaining the equilibrium between mitochondrial Zn^{2+} ($[Zn^{2+}]_m$) and cytosolic Zn^{2+} ($[Zn^{2+}]_c$) is essential for regulating various processes, such as metabolic pathways and cellular signaling [86,87]. While numerous DNAzyme-based technologies have been devised to visualize metal ions in living cells [88–90], methodologies enabling in situ detection specifically within mitochondria remain underdeveloped, hindering further advancements in this area.

Li's group presents a DNAzyme-based nanodevice tailored for the accurate imaging of subcellular metal ions within mitochondria [83]. As depicted in Figure 4B, the nanodevice, consists of two primary components: a UV light-responsive DNAzyme sensor probe (L-DZ) and a functionalized upconversion nanoparticle (UCNP) serving as a near-infrared (NIR) light-driven, organelle-specific delivery mechanism. The L-DZ probe was developed by strategically reconstructing the Zn^{2+}-selective 17E DNAzyme. A blocker sequence was incorporated to form a hairpin structure that prevents the enzyme strand from hybridizing with the left arm of the substrate strand, thereby maintaining the DNAzyme in an inactive conformation. The substrate strand was modified with a Cy5 fluorophore and a black hole quencher (BHQ2) flanking the cleavage site, an adenosine ribonucleotide (rA). This configuration suppresses background fluorescence through Förster resonance energy transfer (FRET). Upon exposure to UV light, the photolysis of the photocleavable (PC) linker disrupts the hairpin structure, enabling the formation of active DNAzyme sensors. This activation transitions the DNAzyme from an OFF state to an ON state. In the presence of Zn^{2+}, the substrate strand is cleaved, releasing a Cy5-labeled fragment, which produces a fluorescence signal indicative of metal-ion detection. To facilitate NIR-controlled activation, UCNPs were integrated into the sensor design. Additionally, the nanodevice was functionalized with triphenylphosphonium (TPP) for targeted mitochondrial delivery of L-DZ. Owing to its lipophilicity and strong positive charge, TPP is widely utilized for directing various cargos to mitochondria, leveraging the organelle's negative membrane potential. The resulting L-DZ/mUC system enables precise visualization of Zn^{2+} within mitochondria by combining controlled localization with remote photoactivation, offering a robust platform for subcellular metal-ion imaging.

Li's group proposed a ribosomal RNA-regulated DNAzyme sensor technology for spatially restricted imaging of Zn^{2+} within mitochondria [84]. As depicted in Figure 4C, the ribosomal RNA-regulated DNAzyme sensor (R-DZ/MB) comprises two components: an rRNA-activatable DNAzyme sensor module (R-DZ) and a molecular beacon-based signal amplification module. The R-DZ is engineered from the Zn^{2+}-specific 17E DNAzyme, incorporating a blocker DNA sequence (B-DNA) with a toehold region [84]. The hybridization between the DNAzyme and B-DNA inhibits the enzyme strand from binding the substrate strand embedded in the MB loop, thus inactivating the DNAzyme's catalytic function. The molecular beacon is designed with a cleavage site (adenosine ribonucleotide, rA) in the loop region and a fluorophore/quencher (Cy5/BHQ2) pair at the terminus of the extended stem, ensuring a low fluorescence background due to Förster resonance energy transfer. Upon addition of rRNA, the B-DNA forms a more stable duplex with rRNA via toehold-mediated strand displacement, releasing the active DNAzyme. In the presence of Zn^{2+}, the DNAzyme cleaves the substrate molecular beacon, breaking it into two segments, leading to MB stem dehybridization and fluorescence recovery for metal ion sensing. The DNAzyme is then liberated from the molecular beacon segments due to reduced hybridization affinity and can participate in subsequent cycles of molecular beacon cleavage, achieving signal amplification via enzymatic multiple turnovers. Then, a mitochondria-targeted nanoparticle (mNP) was developed and combined with the 12S rRNA-responsive R-DZ/MB sensor to create a nanosensor, referred to as R-DZ/MB-mNP (Figure 4D). The sensing capability of R-DZ/MB is specifically activated by the mitochondria-exclusive 12S rRNA following the precise delivery of R-DZ/MB-mNP to the mitochondria, while showing no response to metal ions in regions outside the target. This design enables the nanosensor to achieve spatially specific and amplified imaging of metal ions within subcellular compartments by integrating organelle-targeting localization with rRNA-activated sensing mechanisms. Furthermore, the platform was employed to monitor mitochondrial Zn^{2+} fluctuations in real-time during ischemia and subsequent pharmacological treatment [84].

About overcoming mitochondrial membrane barriers, we concluded that the FNA nanostructures realize this mainly through the combination of membrane potential-driven accumulation, structural engineering, and biofunctional. First, mitochondria maintain a highly negative membrane potential (~-180 mV), which facilitates the accumulation of cationic moieties. FNA nanostructures are often conjugated with lipophilic cations, such as TPP, enabling efficient transport across the outer mitochondrial membrane and subsequent translocation into the inner mitochondrial membrane. Second, compact architectures of FNA nanostructures shield the negatively charged phosphate backbone, reducing electrostatic repulsion. Stimuli-responsive conformational changes, where FNA nanostructures undergo shape transformation or disassembly upon encountering mitochondrial microenvironmental cues, facilitate membrane penetration. Finally, beyond passive accumulation, FNA nanostructures can incorporate mitochondria-specific targeting ligands or peptide-based transporters that facilitate receptor-mediated endocytosis or direct mitochondrial translocation.

3. Therapeutics Based on FNA Target Mitochondria

Mitochondria play a crucial role in various cellular processes, including regulating cell proliferation and apoptosis, transmitting signals within cells, and maintaining the balance of cellular redox states. Disruptions in mitochondrial function can trigger a wide range of disorders, such as cancer, neurodegenerative diseases, cardiovascular conditions, and chronic inflammation [1–3]. Notably, the development of tumors demands significantly elevated ATP levels, which are heavily reliant on mitochondrial function. Consequently, mitochondria have become a promising focal point for the development of targeted cancer therapies [91]. A growing body of evidence suggests that therapeutic strategies aimed at mitochondria can optimize the therapeutic benefit by reducing the necessary drug dosage, overcoming resistance to multiple drugs, preventing tumor recurrence, and limiting metastasis, all while minimizing damage to surrounding healthy tissues [92–94]. In recent years, mitochondria-directed FNA nanostructures have garnered significant attention for their potential to enhance treatment effectiveness and precision. This form of therapy involves the targeted delivery of therapeutic compounds to the mitochondria [95] or directly modulating mitochondrial function to exert therapeutic effects [96–98].

3.1. *FNA Nanostructure-Mediated Mitochondria-Targeted Therapeutics*

Nucleic acid-based therapies, including antisense oligonucleotides (ASOs) and small interfering RNAs (siRNAs), have garnered significant interest for the treatment of a wide range of diseases. However, despite the growing approval of these therapeutic agents, effectively controlling and regulating their activity continues to present a significant challenge.

In 2023, an FNA nanostructure, capable of self-assembling via complementary base pairing of designed single-stranded DNA, was developed by the Dong group [99]. This structure, a tetrahedron, was modified at three of its vertices with triphenylphosphine (TPP), cholesterol, and a functional ASO, forming what is referred to as tetrahedral DNA framework-based nanoparticles (TDFNs) (Figure 5A). These TDFNs have demonstrated the ability to cross the blood–brain barrier (BBB), penetrate neuronal cells, and target mitochondria. Additionally, the TDFNs are capable of recognizing miRNA-34a, generating fluorescence signals for Alzheimer's disease diagnosis, while the incorporated ASO reduces miRNA-34a expression, modulating mitochondrial-associated apoptosis pathways and promoting the survival of neurons (Figure 5A).

The Li group presented a UCNP-based nanostructure for NIR light-controlled, enzymatically triggered regulation of gene expression and combinational tumor therapy [100]. As illustrated in Figure 5B, by integrating ASO with a UCNP-based mitochondria-targeted photodynamic therapy (PDT) platform. This nanostructure was designed to achieve precise, spatiotemporally-controlled gene regulation and combine tumor therapies, as follows: (1) the TPP-functionalized nanostructure ensures targeted localization to the mitochondria; (2) upon exposure to NIR light, the UCNPs emit green light, which excites photosensitizers to generate reactive oxygen species (ROS), leading to mitochondrial damage; (3) the ROS generated during PDT induces the accumulation of mitochondrial APE1, which activates EO, triggering the release of ASO to downregulate ASncmtRNAs in tumor cells, thereby facilitating on-demand gene therapy.

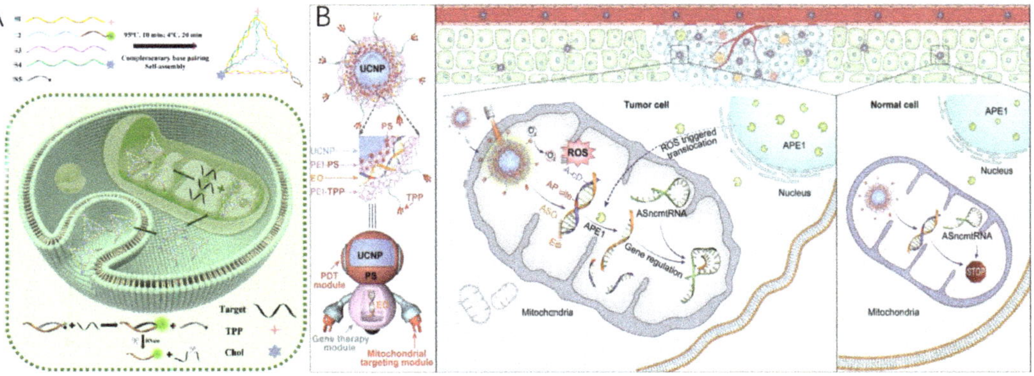

Figure 5. (**A**) Schematic of the composition process of TDFNs and TDFNs for miRNA-34a biomarker detection and AD therapy [99]. (**B**) Schematic of the design and structure of ASO/UCT and its application for spatially selective gene regulation and combinational tumor therapy [100].

3.2. Mitochondria In Situ Self-Assembly of FNA Nanostructure for Therapeutics

The development of dynamically regulated supramolecular assembly systems within living cells to control cell fate represents a novel therapeutic approach [101]. To this end, various materials have been investigated for their ability to enable intracellular self-assembly, thereby modulating cellular functions [102–104]. For instance, nanostructures formed from short peptides have been shown to interfere with mitochondria and hold potential for therapeutic applications. In comparison to peptides, DNA molecules offer remarkable sequence programmability and predictable thermodynamic properties, making them superior candidates for the controlled assembly of dynamic nanostructures [9–12].

In 2023, the Yao group designed an in situ-formed DNA-based network that self-assembled in cancer cells, specifically mediated by telomerase, realizing mitochondrial interference [105]. Two functional DNA modules were designed as depicted in Figure 6A: (i) The Y-shaped DNA (TPP-Y-DNA) was functionalized with TPP to target mitochondria. (ii) The telomerase-responsive linker DNA (T-L-DNA) was constructed with L-DNA and TP to recognize telomerase enzymes. Upon internalization by cancer cells, L-DNA was released from the T-L-DNA via a telomerase enzyme-induced SDR and subsequently hybridized with Y-DNA through base pairing, forming a DNA network. In contrast, this DNA network could not form in normal cells due to the absence of telomerase enzymes. The resulting DNA network encapsulated the mitochondria in cancer cells, disrupting the exchange of substances and inhibiting oxidative phosphorylation and glycolysis, leading to a reduction in ATP production [105]. The diminished ATP levels prevented the formation of the ATP–actin complex, thereby hindering lamellipodium formation, which is crucial for cell migration, and suppressing cancer cell proliferation. Additionally, a significant amount of cytochrome c (Cyt c) was released from the damaged mitochondria into the cytosol, triggering apoptosis in cancer cells.

The Yang group presented a K^+-mediated dynamic assembly of DNA tetrahedrons for mitochondrial interference within living cells. One vertex of the DNA tetrahedron was functionalized with TPP for mitochondria targeting, while the other three vertices were conjugated with guanine-rich sequences to facilitate K^+-induced aggregation of the tetrahedrons (Figure 6B) [106]. These DNA aggregates specifically localized to the mitochondria via TPP and disrupted mitochondrial functions, leading to a marked inhibition of cell migration. Investigation into the underlying mechanism revealed that the negatively charged DNA aggregates acted as a physical barrier on the mitochondria, obstructing substance exchange and consequently suppressing both aerobic respiration and glycolysis. This reduction in ATP production impaired the formation of the ATP–actin complex, thus inhibiting the development of lamellipodia, which are crucial for cellular movement.

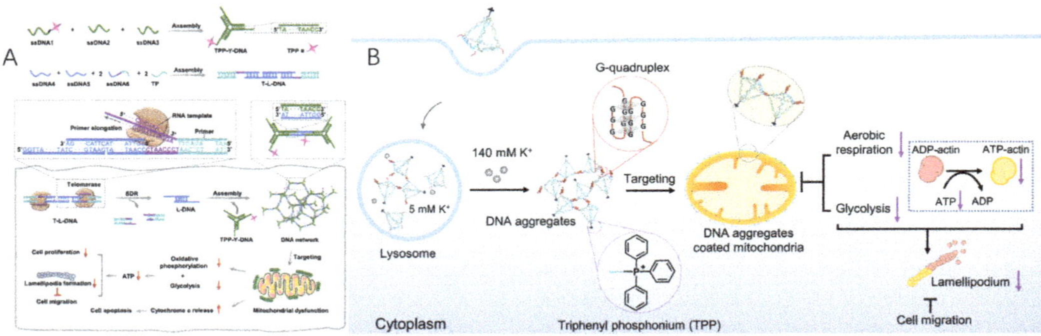

Figure 6. (**A**) Schematic of the assembly of TPP-Y-DNA and T-L-DNA and telomerase-mediated self-assembly of the DNA network inside cancer cells enabling mitochondrial interference and regulation of cellular behaviors [105]. The self-assembly of a DNA network within cancer cells, facilitated by telomerase activity, enables mitochondrial disruption and influences cellular functions. Telomerase, a distinctive ribonucleoprotein enzyme, is responsible for elongating telomeric DNA and extending telomeric primers. When telomerase is present, it catalyzes the primer extension in T-L-DNA, initiating strand displacement reactions and subsequently releasing L-DNA. The liberated L-DNA then hybridizes with TPP-Y-DNA through complementary base pairing, leading to the formation of an organized DNA network. (**B**) Schematic of the dynamic assembly of TDNs-G-TPP in living cells for mitochondrial interference and the consequent regulation of cellular behaviors [106].

4. Conclusions and Perspectives

The mentioned strategies based on FNA nanostructures targeting mitochondria are summarized in Table 1. As shown in Table 1, the strategies' efficiency, biostability, off-target effects, and disadvantages are listed and compared. In conclusion, almost all FNA nanostructure-targeting strategies have high targeting efficiency and relatively good biostability. However, in some strategies, the off-target effect was not assessed by researchers even though it is actually a very critical indicator for evaluating the final targeting capability. By analyzing the disadvantages of these targeting strategies in Table 1, we conclude that the complex construction process is a common disadvantage of almost all targeting strategies based on FNA nanostructures. Meanwhile, some targeting strategies were triggered by light, but the light penetration depth in tissue is still in doubt.

This review summarizes and discusses recent progress in mitochondria-targeted molecular imaging and therapeutics facilitated by FNA nanostructures. By combining mitochondria-targeted nanodelivery approaches with various FNA-based sensing systems, spatially selective imaging of mitochondria-specific analytes has become achievable. For therapeutic applications, two primary strategies were highlighted. The first strategy involves the design of FNA nanodevices for the activatable delivery of therapeutic agents to mitochondria. For example, by combining pre-blocked ASO with a mitochondria-targeted photodynamic therapy (PDT) system, enzyme-activated gene regulation and combinatorial cancer therapy were achieved. Additionally, by tagging photosensitizers (PSs) and quenchers to the trans-cleavage substrate of the CRISPR-Cas12a system, PDT could be selectively triggered in the mitochondria of tumor cells with characteristic mtDNA mutations, further activating the cGAS–STING pathway for immunotherapy. The second strategy involves the use of conditionally controlled in situ FNA assembly to directly interfere with mitochondrial function in tumor cells.

Table 1. Efficiency, biostability, off-target effect, and disadvantages of FNA nanostructure-based mitochondria-targeting strategies.

ategies	Efficiency	Biostability	Off-Target Effect	Disadvantages
NS-hairpin DNA for RNA	High	Excellent	-	Dependence on dual-color imaging analysis instruments; without in vivo experiments
F-Cas12a/crRNA for RNA	High	Good	-	Off-target effect was not discussed
-HAPT for APE1	High	Excellent	Low	Limitation of light penetration depth in tissue; complex construction process
G/NT for ATP and GSH	High	Good	Low	Complex construction process
Z/MUC for Zn^{2+}	High	Good	Low	Complex construction process; without in vivo experiments
Z/MB for Zn^{2+}	High	Doubtful	-	Complex construction process; without in vivo experiments; signal attenuation in deep tissues in vivo
FNs for miRNA-34a	High	Excellent	Doubtful	Complex construction process
/UCT for gene regulation	High	Excellent	Low	Limitation of light penetration depth in tissue; complex construction process
A network for ATP	High	Excellent	Low	Telomerase expression level differences affect the applicability of DNA network
Ns-G-TPP	High	Excellent	Low	Affected by K^+ concentration level differences

Despite significant advancements in recent years, the field of mitochondria-targeted biosensing and therapy remains in its early stages, with numerous challenges yet to be addressed. The success of both mitochondrial sensing and therapeutic strategies heavily relies on the efficient delivery of engineered DNA components. Therefore, developing more advanced targeting approaches for the precise and effective delivery of functional DNA units to mitochondria continues to be a critical focus. For instance, lipid nanoparticles, commonly used in mRNA vaccines, could be engineered with mitochondria-targeting ligands to enhance the delivery of DNA systems [107]. TPP is currently the most widely used ligand for mitochondrial targeting, but other delocalized lipophilic cations, such as dequalinium, guanidine, and biguanide, offer additional possibilities for constructing mitochondria-targeted FNA nanostructures [108,109]. Furthermore, mitochondria-penetrating peptides (MPPs) and mitochondria-targeting sequences (MTSs) have not been fully explored in FNA nanostructure engineering, largely due to limitations in suitable modification methods. Peptide nucleic acids (PNAs) may serve as a promising solution, bridging this gap and facilitating the incorporation of MPPs and MTSs into mitochondria-targeted FNA nanostructures. Additionally, DNA origami nanostructures have emerged as a highly programmable and versatile platform for mitochondrial targeting, offering unprecedented precision in drug delivery, gene regulation, and biomolecular interactions. By leveraging the structural programmability and high cargo-loading capacity of DNA origami, researchers have developed mitochondria-targeting systems that can navigate the complex intracellular environment and overcome mitochondrial membrane barriers [110]. These nanostructures also can be functionalized with TPP to facilitate selective accumulation within mitochondria, thereby enhancing therapeutic efficacy while minimizing off-target effects. Meanwhile, some DNA origami platforms also can be used to study lipid transport between bilayers, which is relevant to mitochondrial and inter-organelle communication [111,112].

Mitochondria-targeted FNA nanostructures are designed to specifically localize in mitochondria and perform molecular imaging functions. However, these nanostructures may become irreversibly activated upon encountering their targets, leading to sensing before actual targeting occurs. To overcome this limitation, the light-activatable sensing approach was used for the programmable response of FNA nanostructures. Nevertheless, the light activation strategy still faces challenges due to the limited subcellular resolution of light. Given that several mitochondria-specific components have been identified, we hypothesize that these components could serve as endogenous mitochondrial triggers to regulate pre-blocked DNA probes, enabling precise subcellular imaging. For example, mtDNA sequences, which are uniquely localized in mitochondria, could act as stimuli to activate the sensing function of mitochondria-targeted biosensors. Additionally, most of the current mitochondria-targeted sensing methods have been validated primarily with well-known molecules. We are optimistic that these strategies can be effectively applied to uncover previously unexplored biological processes within mitochondria and in mitochondrial–organelle communications.

When applying FNA nanostructures for gene editing and mitochondria-targeted tumor or Alzheimer's disease therapies, there is a potential risk of unintended effects on normal cells. Therefore, enhancing the target cell specificity of these mitochondria-targeted FNA systems is crucial. To address this challenge, intelligent FNA nanomachines, capable of responding and adapting to complex biological signals, hold promise as next-generation tools for mitochondria-targeted treatments. These smart FNA nanodevices can be meticulously engineered to incorporate cascade targeting mechanisms, thereby increasing targeting accuracy. For instance, DNA nanodevices can be equipped with tumor-specific ligands, such as aptamers that bind to overexpressed membrane proteins on tumor cells. Upon internalization into target cells, these devices could be directed to mitochondrial compartments using organelle-targeting ligands, allowing them to perform their therapeutic functions. Moreover, spatiotemporal control over the therapeutic actions of mitochondria-targeted systems provides an effective strategy for minimizing off-target effects. In this approach, the therapeutic activity of the DNA nanodevices is initially suppressed and later activated by tumor-specific triggers within mitochondria (such as reactive oxygen species or abnormal metabolite accumulation), thereby enhancing the selective targeting of cancer cells and reducing cytotoxicity to normal tissues. Additionally, with numerous ligands available for targeting different organelles, FNA nanostructures can be engineered to also target other cellular compartments, such as the nucleus or endoplasmic reticulum, for subcellular-specific imaging and therapies. Moving forward, more focus should be placed on designing intelligent FNA nanodevices that integrate both disease-specific recognition and organelle-targeting capabilities. Ultimately, we are optimistic that advancements in organelle-targeted FNA nanostructures will significantly contribute to both academic research and clinical applications.

Funding: This research was funded by Key R&D Program of Shandong Province, China grant number 2022TZXD0033, Science and Technology SMEs Innovation Ability Improvement Project of Shandong Province grant numbers 2023TSGC0386 and 2023TSGC0362, Doctoral Research Startup Foundation of Liaocheng University grant number 318052315 and 318052366. The APC was funded by Qinghua Zeng.

Institutional Review Board Statement: Not applicable.

Informed Consent Statement: Not applicable.

Data Availability Statement: Data will be made available on request.

Conflicts of Interest: The authors declare no conflicts of interest.

ferences

Deus, C.M.; Yambire, K.F.; Oliveira, P.J.; Raimundo, N. Mitochondria-lysosome crosstalk: From physiology to neurodegeneration. *Trends Mol. Med.* **2020**, *26*, 71–88. [CrossRef] [PubMed]

L., S.; Song, M.; Li, Q.; Kirk, A.; Xu, H. Mitochondrion-endothelial interaction generates danger signal in promoting costimulation-resistant T cell-mediated allograft rejection. *Am. J. Transplant.* **2020**, *20*, 335.

Wang, X.R.; Chen, Q.; Ou, J.L.; Huang, Y.L.; Wang, C.C.; Chen, Y.A.; Qin, J.L.; Chen, M.; Feng, Y.; Xu, G.Y.; et al. Real-time and accurate monitoring of mitochondria-related apoptosis by a multifunctional two-photon fluorescent probe. *Sens. Actuators B Chem.* **2022**, *351*, 130953. [CrossRef]

Teulière, J.; Bernard, G.; Bapteste, E. The distribution of genes associated with regulated cell death is decoupled from the mitochondrial phenotypes within unicellular eukaryotic hosts. *Front. Cell Dev. Biol.* **2020**, *8*, 536389. [CrossRef]

Poór, P. Effects of salicylic acid on the metabolism of mitochondrial reactive oxygen species in plants. *Biomolecules* **2020**, *10*, 341. [CrossRef]

Martin-Guerrero, S.M.; Markovinovic, A.; Morotz, G.M.; Salam, S.; Noble, W.; Miller, C.C.J. Targeting ER-mitochondria signaling as a therapeutic target for frontotemporal dementia and related amyotrophic lateral sclerosis. *Front. Cell Dev. Biol.* **2022**, *10*, 915931. [CrossRef]

Yaqoob, M.D.; Xu, L.; Li, C.F.; Leong, M.M.L.; Xu, D.D. Targeting mitochondria for cancer photodynamic therapy. *Photodiagnosis Photodyn. Ther.* **2022**, *38*, 102830. [CrossRef]

Huang, Y.Y.; Zhang, G.X.; Zhao, R.; Zhang, D.Q. Aggregation-induced emission luminogens for mitochondria-targeted cancer therapy. *ChemMedChem* **2020**, *15*, 2220–2227. [CrossRef]

Xu, W.T.; He, W.C.; Du, Z.H.; Zhu, L.Y.; Huang, K.L.; Lu, Y.; Luo, Y.B. Functional nucleic acid nanomaterials: Development, properties, and applications. *Angew. Chem. Int. Ed.* **2021**, *60*, 6890–6918. [CrossRef]

Bialy, R.M.; Mainguy, A.; Li, Y.F.; Brennan, J.D. Functional nucleic acid biosensors utilizing rolling circle amplification. *Chem. Soc. Rev.* **2022**, *51*, 9009–9067. [CrossRef]

Cao, X.; Chen, C.; Zhu, Q. Biosensors based on functional nucleic acids and isothermal amplification techniques. *Talanta* **2023**, *253*, 123977. [CrossRef] [PubMed]

Yu, X.H.; Zhang, S.Q.; Guo, W.Q.; Li, B.X.; Yang, Y.; Xie, B.Q.; Li, K.; Zhang, L. Recent advances on functional nucleic-acid biosensors. *Sensors* **2021**, *21*, 7109. [CrossRef] [PubMed]

Su, M.; Peng, T.; Zhu, Y.; Li, J. Nucleic acid covalent tags. *ChemBioChem* **2024**, e202400805. [CrossRef]

Zhou, W.H.; Saran, R.; Liu, J.W. Metal sensing by DNA. *Chem. Rev.* **2017**, *117*, 8272–8325. [CrossRef] [PubMed]

Herrera, A.; Zhou, J.; Song, M.-S.; Rossi, J.J. Evolution of cell-type-specific RNA aptamers via live cell-based SELEX. *Methods Mol. Biol.* **2023**, *2666*, 317–346.

Sun, P.; Gou, H.; Che, X.; Chen, G.; Feng, C. Recent advances in DNAzymes for bioimaging, biosensing and cancer therapy. *Chem. Commun.* **2024**, *60*, 10805–10821. [CrossRef]

Wang, D.; Li, S.H.; Zhao, Z.L.; Zhang, X.B.; Tan, W.H. Engineering a second-order DNA logic-gated nanorobot to sense and release on live cell membranes for multiplexed diagnosis and synergistic therapy. *Angew. Chem. Int. Ed.* **2021**, *60*, 15816–15820. [CrossRef]

Zheng, K.; Kros, J.M.; Li, J.; Zheng, P.P. DNA-nanorobot-guided thrombin-inducing tumor infarction: Raising new potential clinical concerns. *Drug Discov. Today* **2020**, *25*, 951–955. [CrossRef]

Chen, C.; Wang, X.J.; Wang, Y.; Tian, L.F.; Cao, J.X. Construction of protocell-based artificial signal transduction pathways. *Chem. Commun.* **2021**, *57*, 12754–12763. [CrossRef]

Xie, M.; Jiang, J.; Chao, J. DNA-based gold nanoparticle assemblies: From structure constructions to sensing applications. *Sensors* **2023**, *23*, 9229. [CrossRef]

Zhang, Q.; Gao, L.; Li, F.; Bi, Y. Sensing and manipulating single lipid vesicles using dynamic DNA nanotechnology. *Nanoscale* **2023**, *15*, 5158–5166. [CrossRef] [PubMed]

van Dongen, J.E.; Eijkel, J.C.T.; Segerink, L.I. DNA-origami enabled distance-dependent sensing. In Proceedings of the 2022 IEEE Sensors, Dallas, TX, USA, 8 December 2022; IEEE: Piscataway, NJ, USA, 2022.

Liu, Z.H.; Cao, Y.J.; Zhang, X.N.; Yang, H.Z.; Zhao, Y.J.; Gao, W.; Tang, B. A dual-targeted CeO_2-DNA nanosensor for real-time imaging of H_2O_2 to assess atherosclerotic plaque vulnerability. *J. Mater. Chem. B* **2020**, *8*, 3502–3505. [CrossRef] [PubMed]

Lee, S.; Batjikh, I.; Kang, S.H. Toward sub-diffraction imaging of single-DNA molecule sensors based on stochastic switching localization microscopy. *Sensors* **2020**, *20*, 6657. [CrossRef] [PubMed]

25. Arnon, Z.A.; Piperno, S.; Redeker, D.C.; Randall, E.; Tkachenko, A.V.; Shpaisman, H.; Gang, O. Acoustically shaped D programmable materials. *Nat. Commun.* **2024**, *15*, 6875. [CrossRef]
26. Wang, H.G.; Qi, H.C.; Gong, S.J.; Huang, Z.H.; Meng, C.; Zhang, Y.; Chen, X.; Jiao, X. Fe_3O_4 composited with MoS_2 bl horizontal gene transfer. *Colloids Surf. B. Biointerfaces* **2020**, *185*, 110569. [CrossRef]
27. Maity, S.; Guchhait, R.; De, S.; Pramanick, K. High doses of nano-polystyrene aggravate the oxidative stress, DNA damage, the cell death in onions. *Environ. Pollut.* **2023**, *316 Pt 2*, 120611. [CrossRef]
28. Hu, M.; Palic, D. Role of microRNAs in regulation of DNA damage in monocytes exposed to polystyrene and TiO_2 nanopart *Toxicol. Rep.* **2020**, *7*, 743–751. [CrossRef]
29. Huang, Q.; Liu, X.; Zhang, P.G.; Wu, Z.; Zhao, Z.L. A DNA nano-train carrying a predefined drug combination for cancer the *Chem. Res. Chin. Univ.* **2022**, *38*, 928–934. [CrossRef]
30. Narita, M.; Kohata, A.; Kageyama, T.; Watanabe, H.; Aikawa, K.; Kawaguchi, D.; Morihiro, K.; Okamoto, A.; Okazc Fluorocarbon-DNA conjugates for enhanced cellular delivery: Formation of a densely packed DNA nano-assembly. *ChemBioC* **2024**, *25*, e202400436. [CrossRef]
31. Liu, Y.; Lin, Y.; Xiao, H.; Fu, Z.; Zhu, X.; Chen, X.; Li, C.; Ding, C.; Lu, C. mRNA-responsive two-in-one nanodrug for enha anti-tumor chemo-gene therapy. *J. Control. Release* **2024**, *369*, 765–774. [CrossRef]
32. Rajwar, A.; Shetty, S.R.; Vaswani, P.; Morya, V.; Barai, A.; Sen, S.; Sonawane, M.; Bhatia, D. Geometry of a DNA nanostruc influences its endocytosis: Cellular study on 2D, 3D, and in vivo systems. *ACS Nano* **2022**, *16*, 10496–10508. [CrossRef] [PubM
33. Zhang, A.H.; Fang, J.R.; Li, X.L.; Wang, J.; Chen, M.W.; Chen, H.J.; He, G.; Xie, X. Cellular nanointerface of vertical nanostru arrays and its applications. *Nanoscale Adv.* **2022**, *4*, 1844–1867. [CrossRef] [PubMed]
34. Haldavnekar, R.; Venkatakrishnan, K.; Tan, B. Boosting the sub-cellular biomolecular cancer signals by self-functionalized tag nano sensor. *Biosens. Bioelectron.* **2021**, *190*, 113407. [CrossRef] [PubMed]
35. Saminathan, A.; Devany, J.; Veetil, A.T.; Suresh, B.; Pillai, K.S.; Schwake, M.; Krishnan, Y. A DNA-based voltmeter for organ *Nat. Nanotechnol.* **2021**, *16*, 96–103. [CrossRef]
36. Iwaki, M.; Wickham, S.F.; Ikezaki, K.; Yanagida, T.; Shih, W.M. A programmable DNA origami nanospring that reveals f induced adjacent binding of myosin VI heads. *Nat. Commun.* **2016**, *7*, 13715. [CrossRef]
37. Lin, M.; Chen, Y.; Zhao, S.; Tang, R.; Nie, Z.; Xing, H. A biomimetic approach for spatially-controlled cell membrane engine using fusogenic spherical nucleic acid. *Angew. Chem. Int. Ed.* **2021**, *61*, e202111647. [CrossRef]
38. Basu, U.; Bostwick, A.M.; Das, K.; Dittenhafer-Reed, K.E.; Patel, S.S. Structure, mechanism, and regulation of mitochondrial transcription initiation. *J. Biol. Chem.* **2020**, *295*, 18406–18425. [CrossRef]
39. Esfandyari, D.; Idrissou, B.M.G.; Hennis, K.; Avramopoulos, P.; Dueck, A.; El-Battrawy, I.; Grüter, L.; Meier, M.A.; Näger, Ramanujam, D.; et al. MicroRNA-365 regulates human cardiac action potential duration. *Nat. Commun.* **2022**, *13*, 220. [Cros
40. Trabucchi, M. Subcellular heterogeneity of the microRNA machinery. *Trends Genet.* **2019**, *35*, 15–28. [CrossRef]
41. Park, A.; Oh, M.; Lee, S.J.; Oh, K.J.; Lee, E.W.; Lee, S.C.; Bae, K.H.; Han, B.S.; Kim, W.K. Mitochondrial transplantation as a r therapeutic strategy for mitochondrial diseases. *Int. J. Mol. Sci.* **2021**, *22*, 4793. [CrossRef]
42. Bonora, M.; Wieckowski, M.R.; Sinclair, D.A.; Kroemer, G.; Pinton, P.; Galluzzi, L. Targeting mitochondria for cardiovas disorders: Therapeutic potential and obstacles. *Nat. Rev. Cardiol.* **2019**, *16*, 33–55. [CrossRef] [PubMed]
43. Ortega, M.A.; Fraile-Martínez, O.; Guijarro, L.G.; Casanova, C.; Coca, S.; Alvarez-Mon, M.; Buján, J.; García-Honduvill Asúnsolo, A. The regulatory role of mitochondrial microRNAs (mitomiRs) in breast cancer: Translational implications pr and future. *Cancers* **2020**, *12*, 2443. [CrossRef] [PubMed]
44. Cheng, Y.-Y.; Chen, Z.; Cao, X.; Ross, T.D.; Falbel, T.G.; Burton, B.M.; Venturelli, O.S. Programming bacteria for multiplexed detection. *Nat. Commun.* **2023**, *14*, 2001. [CrossRef] [PubMed]
45. Wang, Y.; Wang, D.; Sun, L.H.; Zhang, L.C.; Lu, Z.S.; Xue, P.; Wang, F.; Xia, Q.Y.; Bao, S.J. BC@DNA-$Mn_3(PO_4)_2$ nanozym real-time detection of superoxide from living cells. *Anal. Chem.* **2020**, *92*, 15927–15935. [CrossRef]
46. Taghbalout, A.; Jillette, N.; Cheng, A.W. TALE.sense: A versatile DNA sensor platform for live mammalian cells. *ACS Synth* **2022**, *11*, 116–124. [CrossRef]
47. Shapiro, J.A. All living cells are cognitive. *Biochem. Biophys. Res. Commun.* **2021**, *564*, 134–149. [CrossRef]
48. Chen, Z.; Zeng, S.; Qian, L. Quantitative analysis of mitochondrial RNA in living cells with a dual-color imaging system. **2023**, *19*, e2301132. [CrossRef]
49. Chen, W.S.; Ouyang, J.; Liu, H.; Chen, M.; Zeng, K.; Sheng, J.P.; Liu, Z.J.; Han, Y.J.; Wang, L.Q.; Li, J.; et al. Black phosph nanosheet-based drug delivery system for synergistic photodynamic/photothermal/chemotherapy of cancer. *Adv. Mater.* *29*, 1603864. [CrossRef]
50. Chen, W.S.; Ouyang, J.; Yi, X.Y.; Xu, Y.; Niu, C.C.; Zhang, W.Y.; Wang, L.Q.; Sheng, J.P.; Deng, L.; Liu, Y.N.; et al. Black phosp nanosheets as a neuroprotective nanomedicine for neurodegenerative disorder therapy. *Adv. Mater.* **2018**, *30*, 1703458. [Cros
51. Shao, J.D.; Xie, H.H.; Huang, H.; Li, Z.B.; Sun, Z.B.; Xu, Y.H.; Xiao, Q.L.; Yu, X.F.; Zhao, Y.T.; Zhang, H.; et al. Biodegradable phosphorus-based nanospheres for in vivo photothermal cancer therapy. *Nat. Commun.* **2021**, *12*, 3923. [CrossRef]

Chen, H.L.; Liu, Z.M.; Wei, B.; Huang, J.; You, X.R.; Zhang, J.Y.; Yuan, Z.L.; Tang, Z.L.; Guo, Z.Y.; Wu, J. Redox responsive nanoparticle encapsulating black phosphorus quantum dots for cancer theranostics. *Bioact. Mater.* **2021**, *6*, 655–665. [CrossRef] [PubMed]

Liu, W.X.; Dong, A.; Wang, B.; Zhang, H. Current advances in black phosphorus-based drug delivery systems for cancer therapy. *Adv. Sci.* **2021**, *8*, 2003033. [CrossRef] [PubMed]

Xu, X.; Jiang, Y.F.; Wang, M.; Wang, H.M.; Lu, C.H.; Yang, H.H. Biodegradable black-phosphorus-nanosheet-based nanoagent for enhanced chemo-photothermal therapy. *Part. Part. Syst. Charact.* **2020**, *37*, 2000243. [CrossRef]

Huang, W.-Q.; Wang, F.; Nie, X.; Zhang, Z.; Chen, G.; Xia, L.; Wang, L.-H.; Ding, S.-G.; Hao, Z.-Y.; Zhang, W.-J.; et al. Stable black phosphorus nanosheets exhibiting high tumor-accumulating and mitochondria-targeting for efficient photothermal therapy via double functionalization. *ACS Appl. Bio Mater.* **2020**, *3*, 1176–1186. [CrossRef]

Zhang, Q.C.; Liang, J.M.; Hu, X.W.; Cai, A.; Zhu, Y.Z.; Peng, W.C.; Li, Y.; Zhang, F.B.; Fan, X.B. Rapid microwave-assisted synthesis of Ni_2P nanosheets from black phosphorus. *Chem. Commun.* **2022**, *58*, 10945–10948. [CrossRef]

Hou, Y.B.; Xu, Z.M.; An, R.; Zheng, H.; Hu, W.Z.; Zhou, K. Recent progress in black phosphorus nanosheets for improving the fire safety of polymer nanocomposites. *Compos. Part B Eng.* **2023**, *249*, 110404. [CrossRef]

Wen, Y.; Liu, W.-Y.; Wang, J.-H.; Yu, Y.-L.; Chen, S. Simultaneous imaging of multiple miRNAs in mitochondria controlled by fluorescently encoded upconversion optical switches for drug resistance studies. *Anal. Chem.* **2023**, *95*, 12152–12160. [CrossRef]

Li, L.L.; Cheng, F.; Li, J.; Zhen, S.J.; Lv, W.Y.; Shuai, X.J.; Li, Y.F.; Huang, C.Z.; Li, C.M. Simultaneous response of mitochondrial microRNAs to identify cell apoptosis with multiple responsive intelligent DNA biocomputing nanodevices. *Anal. Chem.* **2023**, *95*, 10992–10998. [CrossRef]

Du, J.Y.; Qiao, Y.C.; Meng, X.D.; Wei, W.; Dai, W.H.; Yang, L.Z.; Yang, C.Y.; Dong, H.F. Mitochondria microRNA spatial imaging via pH-responsive exonuclease-assisted AIE nanoreporter. *Anal. Chem.* **2022**, *94*, 10669–10675. [CrossRef]

Li, Y.; Wu, Y.; Xu, R.; Guo, J.; Quan, F.; Zhang, Y.; Huang, D.; Pei, Y.; Gao, H.; Liu, W.; et al. In vivo imaging of mitochondrial DNA mutations using an integrated nano Cas12a sensor. *Nat. Commun.* **2023**, *14*, 7722. [CrossRef]

Wang, B.; Lv, X.J.; Wang, Y.F.; Wang, Z.B.; Liu, Q.; Lu, B.; Liu, Y.; Gu, F. CRISPR/Cas9-mediated mutagenesis at microhomologous regions of human mitochondrial genome. *Sci. China Life Sci.* **2021**, *64*, 1463–1472. [CrossRef] [PubMed]

Wang, Y.; Hu, L.F.; Cui, P.F.; Qi, L.Y.; Xing, L.; Jiang, H.L. Pathologically responsive mitochondrial gene therapy in an allotopic expression-independent manner cures leber's hereditary optic neuropathy. *Adv. Mater.* **2021**, *33*, 2103307. [CrossRef] [PubMed]

Yuan, P.Y.; Mao, X.; Wu, X.F.; Liew, S.S.; Li, L.; Yao, S.Q. Mitochondria-targeting, intracellular delivery of native proteins using biodegradable silica nanoparticles. *Angew. Chem. Int. Ed.* **2019**, *58*, 7657–7661. [CrossRef] [PubMed]

Zielonka, J.; Joseph, J.; Sikora, A.; Hardy, M.; Ouari, O.; Vasquez-Vivar, J.; Cheng, G.; Lopez, M.; Kalyanaraman, B. Mitochondria-targeted triphenylphosphonium-based compounds: Syntheses, mechanisms of action, and therapeutic and diagnostic applications. *Chem. Rev.* **2017**, *117*, 10043–10120. [CrossRef]

Lee-Glover, L.P.; Shutt, T.E. Mitochondrial quality control pathways sense mitochondrial protein import. *Trends Endocrinol. Metab.* **2024**, *35*, 308–320. [CrossRef]

Wallis, C.P.; Scott, L.H.; Filipovska, A.; Rackham, O. Manipulating and elucidating mitochondrial gene expression with engineered proteins. *Philos. Trans. R. Soc. B Biol. Sci.* **2020**, *375*, 20190185. [CrossRef]

Mol, C.D.; Izumi, T.; Mitra, S.; Tainer, J.A. DNA-bound structures and mutants reveal abasic DNA binding by APE1 and DNA repair coordination. *Nature* **2000**, *403*, 451–456. [CrossRef]

Lebedeva, N.A.; Dyrkheeva, N.S.; Rechkunova, N.I.; Lavrik, O.I. Apurinic/apyrimidinic endonuclease 1 has major impact in prevention of suicidal covalent DNA-protein crosslink with apurinic/apyrimidinic site in cellular extracts. *IUBMB Life* **2024**, *76*, 987–996. [CrossRef]

Mahjabeen, I.; Baig, R.M.; Sabir, M.; Kayani, M.A. Genetic and expressional variations of *APEX1* are associated with increased risk of head and neck cancer. *Mutagenesis* **2013**, *28*, 213–218. [CrossRef]

Dan, K.; Veetil, A.T.; Chakraborty, K.; Krishnan, Y. DNA nanodevices map enzymatic activity in organelles. *Nat. Nanotechnol.* **2019**, *14*, 252–259. [CrossRef]

Lai, R.; Pan, X.; Qin, Y.; Liang, J.; Wu, L.; Dong, M.; Chen, J.; Liu, J.-W. Highly integrated, self-powered and activatable bipedal DNA nanowalker for imaging of base excision repair in living cells. *J. Nanobiotechnol.* **2024**, *22*, 636. [CrossRef] [PubMed]

Shao, Y.L.; Zhao, J.; Yuan, J.Y.; Zhao, Y.L.; Li, L.L. Organelle-specific photoactivation of DNA nanosensors for precise profiling of subcellular enzymatic activity. *Angew. Chem. Int. Ed.* **2021**, *60*, 8923–8931. [CrossRef] [PubMed]

Cadenas, E.; Davies, K.J. Mitochondrial free radical generation, oxidative stress, and aging. *Free Radic. Biol. Med.* **2000**, *29*, 222–230. [CrossRef] [PubMed]

Droge, W. Free radicals in the physiological control of cell function. *Physiol. Rev.* **2002**, *82*, 47–95. [CrossRef]

Gallez, B.; Mathieu, B.; Sonveaux, P. About metformin and its action on the mitochondrial respiratory chain in prostate cancer. *Transl. Androl. Urol.* **2024**, *13*, 909–914. [CrossRef]

77. Chen, T.-H.; Wang, H.-C.; Chang, C.-J.; Lee, S.-Y. Mitochondrial glutathione in cellular redox homeostasis and disease mani tion. *Int. J. Mol. Sci.* **2024**, *25*, 1314. [CrossRef]
78. Hu, W.W.; Zhang, C.; Wu, R.; Sun, Y.; Levine, A.; Feng, Z.H. Glutaminase 2, a novel p53 target gene regulating energy metab and antioxidant function. *Proc. Natl. Acad. Sci. USA* **2010**, *107*, 7455–7460. [CrossRef]
79. Ali, A.S.; Chen, R.; Raju, R.; Kshirsagar, R.; Gilbert, A.; Zang, L.; Karger, B.L.; Ivanov, A.R. Multi-omics reveals imp cysteine feed concentration and resulting redox imbalance on cellular energy metabolism and specific productivity in CH bioprocessing. *Biotechnol. J.* **2020**, *15*, 1900565. [CrossRef]
80. Hong, S.N.; Zhang, X.T.; Lake, R.J.; Pawel, G.T.; Guo, Z.J.; Pei, R.J.; Lu, Y. A photo-regulated aptamer sensor for spatiotemp controlled monitoring of ATP in the mitochondria of living cells. *Chem. Sci.* **2020**, *11*, 713–720. [CrossRef]
81. Li, Y.; Zhang, F.; Liu, W.; Shao, M.; Hao, Z.; Zhang, H.; Zhang, R.; Li, X.; Zhang, L. Endogenous glutathione-activated nanoprobes for spatially controllable imaging of microRNA in living cells. *Chem. Commun.* **2023**, *59*, 5431–5434. [CrossRef
82. Chai, X.; Fan, Z.T.; Yu, M.M.; Zhao, J.; Li, L.L. A redox-activatable DNA nanodevice for spatially-selective, AND-gated Im of ATP and glutathione in mitochondria. *Nano Lett.* **2021**, *21*, 10047–10053. [CrossRef] [PubMed]
83. Yi, D.; Zhao, H.; Zhao, J.; Li, L. Modular engineering of DNAzyme-based sensors for spatioselective imaging of metal ic mitochondria. *J. Am. Chem. Soc.* **2023**, *145*, 1678–1685. [CrossRef] [PubMed]
84. Yi, D.; Li, L.A.-O.; Li, M. Subcellular compartment-specific amplified imaging of metal ions via ribosomal RNA-regu DNAzyme sensors. *Angew. Chem. Int. Ed.* **2024**, e202412387.
85. Dang, Z.; Tao, X.-Y.; Guan, Y.; Wu, Z.; Xiong, Y.; Liu, G.; Tian, Y.; Tian, L.-J. Direct visualization and restoration of me ion-induced subcellular ultrastructural remodeling. *ACS Nano* **2023**, *17*, 9069–9081. [CrossRef] [PubMed]
86. Turan, B.; Tuncay, E. The role of labile Zn^{2+} and Zn^{2+}-transporters in the pathophysiology of mitochondria dysfuncti cardiomyocytes. *Mol. Cell. Biochem.* **2021**, *476*, 991–992. [CrossRef]
87. Anson, K.J.; Corbet, G.A.; Palmer, A.E. Zn^{2+} influx activates ERK and Akt signaling pathways. *Proc. Natl. Acad. Sci. USA* *118*, e2015786118. [CrossRef]
88. Hong, C.; Wang, Q.; Chen, Y.Y.; Gao, Y.H.; Shang, J.H.; Weng, X.C.; Liu, X.Q.; Wang, F.A. Intelligent demethylase-d DNAzyme sensor for highly reliable metal-ion imaging in living cells. *Chem. Sci.* **2021**, *12*, 15339–15346. [CrossRef]
89. Wang, W.J.; Satyavolu, N.S.R.; Wu, Z.K.; Zhang, J.R.; Zhu, J.J.; Lu, Y. Near-infrared photothermally activated DNAzyme nanoshells for imaging metal ions in living cells. *Angew. Chem. Int. Ed.* **2017**, *56*, 6798–6802. [CrossRef]
90. Liu, R.; Jiang, D.; Yun, Y.; Feng, Z.; Zheng, F.; Xiang, Y.; Fan, H.; Zhang, J. Photoactivatable engineering of CRISPR/Cas9-ind DNAzyme probe for in situ imaging of nuclear zinc ions. *Angew. Chem. Int. Ed.* **2024**, *63*, e202315536. [CrossRef]
91. Yu, F.; Li, X.; Sheng, C.; Li, L. DNA nanotechnology targeting mitochondria: From subcellular molecular imaging to tailor- therapeutics. *Angew. Chem. Int. Ed.* **2024**, *63*, e202409351. [CrossRef]
92. Huang, M.F.; Xiong, D.H.; Pan, J.; Zhang, Q.; Wang, Y.; Myers, C.R.; Johnson, B.D.; Hardy, M.; Kalyanaraman, B.; Yc Prevention of tumor growth and dissemination by in situ vaccination with mitochondria-targeted atovaquone. *Adv. Sci.* **2** 2101267. [CrossRef] [PubMed]
93. Chen, M.; Xu, H.; Chang, P.; Li, X.; Liu, S.; Xu, L.; Xu, K.; Cheng, G. An intelligent poly aptamer-encoded DNA nanocle tumor site activated mitochondria-targeted photodynamic therapy and MR imaging. *Mater. Today Bio* **2024**, *29*, 101318. [Cro [PubMed]
94. Wang, D.-D.; He, L.; Qi, M.-H.; Zhao, H.-Y.; Yu, A.-X.; Huang, S.-W. Mitochondria-targeting artesunate-rhein conju Linker-modulated cell-permeability, heme-affinity and anticancer activity. *Eur. J. Med. Chem.* **2025**, *282*, 117100. [Cro [PubMed]
95. Luo, Z.; Cao, Y.; Liao, Z.; Gong, N.; Ma, P.; Li, Z.; Lai, X.; Zhang, Y.; Zhu, X.; Li, Z.; et al. Mitochondria-targeted biometallization for photoacoustically visualized photothermal cancer therapy. *ACS Nano* **2024**, *18*, 29667. [CrossRef]
96. Li, L.; Zhang, Y.; Tang, Q.; Wu, C.; Yang, M.; Hu, Y.; Gong, Z.; Shi, L.; Guo, C.; Zeng, Z.; et al. Mitochondria in tumor im surveillance and tumor therapies targeting mitochondria. *Cell. Oncol.* **2024**, *47*, 2031–2047. [CrossRef]
97. Wan, X.; Wang, W.; Zhou, Y.; Ma, X.; Guan, M.; Liu, F.; Chen, S.; Fan, J.-X.; Yan, G.-P. Self-delivery nanoplatform bas amphiphilic apoptosis peptide for precise mitochondria-targeting photothermal therapy. *Mol. Pharm.* **2024**, *21*, 1537 [CrossRef]
98. Chen, F.; Xue, Y.; Zhang, W.; Zhou, H.; Zhou, Z.; Chen, T.; YinWang, E.; Li, H.; Ye, Z.; Gao, J.; et al. The role of mitochonc tumor metastasis and advances in mitochondria-targeted cancer therapy. *Cancer Metastasis Rev.* **2024**, *43*, 1419–1443. [Cro
99. Li, W.; Peng, X.; Mei, X.; Dong, M.; Li, Y.; Dong, H. Multifunctional DNA tetrahedron for Alzheimer's disease mitocho targeted therapy by microRNA regulation. *ACS Appl. Mater. Interfaces* **2023**, *15*, 22977–22984. [CrossRef]
100. Chai, X.; Yi, D.; Sheng, C.; Zhao, J.; Li, L. A remotely controlled nanosystem for spatiotemporally specific gene regulatio combinational tumor therapy. *Angew. Chem. Int. Ed.* **2023**, e202217702.
101. Yang, X.J.; Lu, H.L.; Tao, Y.H.; Zhou, L.C.; Wang, H.M. Spatiotemporal control over chemical assembly in living cells by integ of acid-catalyzed hydrolysis and enzymatic reactions. *Angew. Chem. Int. Ed.* **2021**, *60*, 23797–23804. [CrossRef]

Tang, J.; Liu, J.; Zheng, Q.; Yao, R.; Wang, M. Neuroprotective bioorthogonal catalysis in mitochondria using protein-integrated hydrogen-bonded organic frameworks. *Angew. Chem. Int. Ed.* **2023**, *62*, e202312784. [CrossRef] [PubMed]
Dai, W.B.; Deng, Y.Y.; Chen, X.H.; Huang, Y.; Hu, H.T.; Jin, Q.; Tang, Z.; Ji, J. A mitochondria-targeted supramolecular nanoplatform for peroxynitrite-potentiated oxidative therapy of orthotopic hepatoma. *Biomaterials* **2022**, *290*, 121854. [CrossRef] [PubMed]
Li, M.M.; Song, Y.Q.; Song, N.; Wu, G.Y.; Zhou, H.; Long, T.F.; Zhang, P.C.; Shi, L.Q.; Yu, Z.L. Supramolecular antagonists promote mitochondrial dysfunction. *Nano Lett.* **2021**, *21*, 5730–5737. [CrossRef] [PubMed]
Guo, Y.; Li, S.; Tong, Z.; Tang, J.; Zhang, R.; Lv, Z.; Song, N.; Yang, D.; Yao, C. Telomerase-mediated self-assembly of DNA network in cancer cells enabling mitochondrial interference. *J. Am. Chem. Soc.* **2023**, *145*, 23859–23873. [CrossRef]
Li, F.; Liu, Y.; Dong, Y.; Chu, Y.; Song, N.; Yang, D. Dynamic assembly of DNA nanostructures in living cells for mitochondrial interference. *J. Am. Chem. Soc.* **2022**, *144*, 4667–4677. [CrossRef]
Niu, X.; Zhang, J.; Zhang, J.; Bai, L.; Hu, S.; Zhang, Z.; Bai, M. Lipid nanoparticle-mediated oip5-as1 delivery preserves mitochondrial function in myocardial ischemia/reperfusion injury by inhibiting the p53 pathway. *ACS Appl. Mater. Interfaces* **2024**, *16*, 61565–61582. [CrossRef]
Pawar, A.; Korake, S.; Pawar, A.; Kamble, R. Delocalized lipophilic cation triphenyl phosphonium: Promising molecule for mitochondria targeting. *Curr. Drug Deliv.* **2023**, *20*, 1217–1223. [CrossRef]
Li, X.; Li, Y.; Du, M.; Petrov, Y.V.; Baulin, V.E.; Wang, Y.; Yuan, H.; Zhou, Y.; Li, B. Target-oriented synthesis of triphenylphosphine functionalized carbon dots with negative charge for ROS scavenging and mitochondrial targeting. *ACS Appl. Mater. Interfaces* **2024**, *16*, 28991–29002. [CrossRef]
Luu, M.T.; Berengut, J.F.; Li, J.; Chen, J.-B.; Daljit Singh, J.K.; Coffi Dit Glieze, K.; Turner, M.; Skipper, K.; Meppat, S.; Fowler, H.; et al. Reconfigurable nanomaterials folded from multicomponent chains of DNA origami voxels. *Sci. Robot.* **2024**, *9*, eadp2309. [CrossRef]
Bian, X.; Zhang, Z.; Xiong, Q.; De Camilli, P.; Lin, C. A programmable DNA-origami platform for studying lipid transfer between bilayers. *Nat. Chem. Biol.* **2019**, *15*, 830–837. [CrossRef]
Darley, E.; Singh, J.K.D.; Surace, N.A.; Wickham, S.F.J.; Baker, M.A.B. The fusion of lipid and DNA nanotechnology. *Genes* **2019**, *10*, 1001. [CrossRef]

molecules

Review

Application of the Electrospinning Technique in Electrochemic Biosensors: An Overview

Jie Liu [1], Zhong Dong [2], Ke Huan [2], Zhangchu He [2], Qixian Zhang [3,4,*], Dongmei Deng [2] and Liqiang Luo [2,*]

1 School of Environmental and Chemical Engineering, Shanghai University, Shanghai 200444, China; sakura20030106@163.com
2 College of Sciences, Shanghai University, Shanghai 200444, China; dongzhongdz@163.com (Z.D.); hk8808@shu.edu.cn (K.H.)
3 School of Materials Science and Engineering, Shanghai University, Shanghai 200436, China
4 Shaoxing Institute of Technology, Shanghai University, Shaoxing 312000, China
* Correspondence: qxzhang@ciac.ac.cn (Q.Z.); luck@shu.edu.cn (L.L.); Tel.: +86-21-66132404 (L.L.)

Abstract: Electrospinning is a cost-effective and flexible technology for producing nanofibers with large specific surface areas, functionalized surfaces, and stable structures. In recent years, electrospun nanofibers have attracted more and more attention in electrochemical biosensors due to their excellent morphological and structural properties. This review outlines the principle of electrospinning technology. The strategies of producing nanofibers with different diameters, morphologies, and structures are discussed to understand the regulation rules of nanofiber morphology and structure. The application of electrospun nanofibers in electrochemical biosensors is reviewed in detail. In addition, we look towards the future prospects of electrospinning technology and the challenge of scale production.

Keywords: electrospinning; nanofiber; morphology control; electrochemical sensors; review

1. Introduction

Nanofibers, as one-dimensional nanomaterials, have attracted much attention due to their unique advantages of large specific surface areas, functionalized surfaces, and stable structures. The large specific surface areas contribute to the excellent adsorption performance of nanofibers [1,2]. In particular, nanofibers with suitable pore size distribution can provide a large number of sites to accommodate the high loading of active materials [3].

The electrospinning technique is a process of spinning polymer solutions or melts under a strong electric field. Under the action of the electric field, the spinning fluid expands into a tiny jet that solidifies into a fiber. Electrospinning devices are simple and have low cost. Electrospinning can produce a wide variety of materials, and the process is controllable. These advantages make it one of the most popular techniques for producing polymer nanofibers. In addition, carbon nanofibers can be obtained by carbonized polymer precursors [4]. Electrospinning is also a common technique for producing composite nanofibers. In contrast with the traditional spinning technique, a high-voltage electrostatic field can stretch polymer solution (or melt) into nanofiber, making the diameter of the fiber produced by electrospinning to be as small as one nanometer [5]. Moreover, the technique is simple, cost-effective, and versatile, making it suitable for industrial production, as will be discussed later in the review.

Nowadays, with the development of industry, more and more pollutants of inorganic and organic contaminants have been produced. Therefore, qualitative and quantitative analyses of these pollutants play an important role in environmental protection and food safety [6–9]. Many techniques have been developed such as fluorescence, UV-Vis spectroscopy, mass spectrometry, and electroanalysis [10–14]. Among them, electrochemical sensors have received more and more attention due to their excellent sensitivity, accuracy, wide detection range, easy operation, and low price [15].

In biosensors, electrospun nanofibers can be used as substrate materials or functional components of sensors. As the base material, electrospun nanofibers can provide a large surface area to enhance the adsorption of biomolecules, thus improving the sensitivity and detection limit of the sensors. At the same time, the pore structure of electrospun nanofibers is also conducive to the diffusion and transfer of biomolecules, enhancing the response speed and stability of the sensors. In addition, electrospun nanofibers can also be used as functional components, such as fixed carriers of biomolecules and fixed substrates of biometric molecules. By immobilizing biomolecules or biometric molecules onto electrospun nanofibers, a highly sensitive and selective detection of different biomolecules can be achieved.

In this review, we outline the principle of electrospinning technology, the regulation of nanofiber morphology, and the application of electrospun nanofibers in electrochemical sensors. The challenge and prospect of electrospinning technology are also prospected.

2. The Principle of the Electrospinning Technique

As shown in Figure 1A, a simple electrospinning setup requires four basic components: a high-voltage power supply, pump, nozzle, and collector. The high-voltage power supply provides a strong electric field between the nozzle and collector. The solution is controlled by a pump and aggregates into droplets at the nozzle. Under the influence of the electric field, the shape of the droplet changes and a jet is formed (Figure 1B) [16]. Then, the jet volatilizes rapidly and tapers under the perturbation of the electric field (Figure 1C). Finally, the jet solidifies into fibers and is collected by the collector.

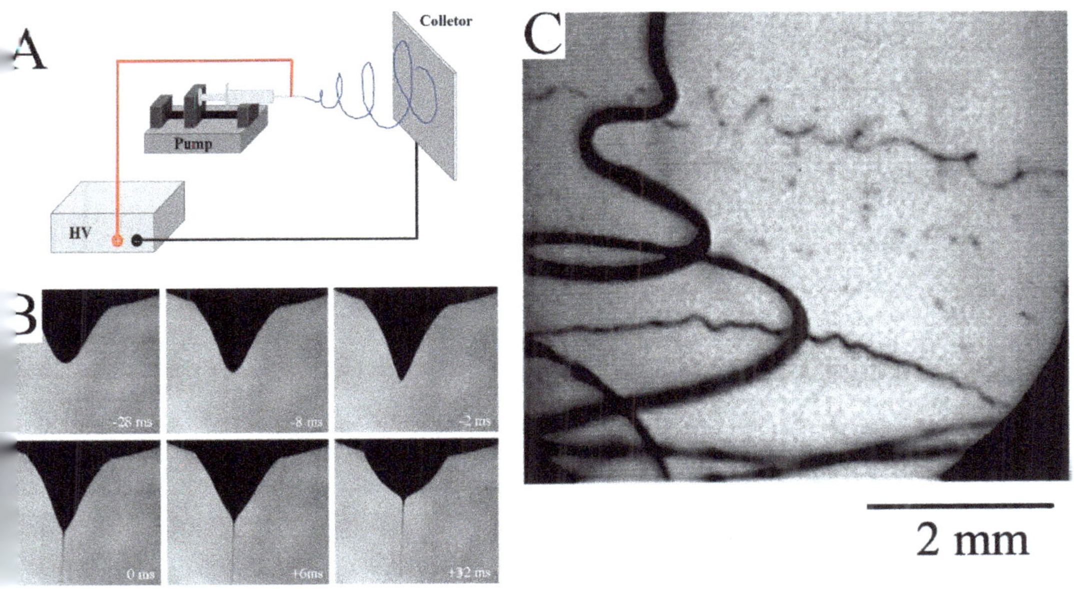

Figure 1. (**A**) Schematic diagram of an electrospinning device. (**B**) The photo shows the process of water droplet deformation, followed by the ejection of a jet. The time at zero was taken to be the frame in which the jet first appeared. (**C**) Image of the jet. The exposure time was 0.25 ms. (**B**) Reprinted with permission from [16]. Copyright 2008, Elsevier. (**C**) Reprinted with permission from [17]. Copyright 2000, American Institute of Physics.

Organic polymers are the most often used materials for electrospinning. If organic polymers do not degrade as they dissolve or melt, they can generally be used directly for electrospinning. Depending on the functions of organic polymers, they can be divided into the following categories: (i) As the host of electrospun nanofibers: The polymers themselves have specific functions. For example, polyvinylidene difluoride (PVDF) nanofibers can be directly used in piezoelectric sensors [18,19]; (ii) As the scaffold or conducting network carrying functional materials: For example, polyacrylonitrile (PAN) nanofibers loaded with Fe_3O_4 nanoparticles were used for vitamin D3 detection [20]; (iii) As a sacrificial phase: For example, after the $AgNO_3/Co(Ac)_2$/polyvinylpyrrolidone (PVP) precursor nanofibers were calcined, Au-Ag/Co_3O_4 nanofibers were prepared, and the PVP was sacrificed in the process [21]; and (iv) As the precursor of carbon nanofibers (CNFs): Some polymers, such as PAN and PVP, were carbonized to form CNFs after thermal treatment in an inert atmosphere [22,23]. In addition, small molecules can also be used for electrospinning under special conditions, and their highly concentrated solutions or pure melts can form self-assembled structures to entangle together and behave like polymer chains [24].

In 1964, Taylor mathematically described and modeled the disintegration of drop in an electric field [25]. As reported, as the strength of the electric field exceeded the critical value, the droplets gradually changed to cones with a half angle of 49.3°. Jets were projected from these cones, which were later named as "Taylor cone". The formation of Taylor cones is caused by the joint influence of electric field and surface tension. Liquid tends to be spherical under the influence of surface tension. The electric field causes the surface of droplets to accumulate large amounts of homogenous charges. As the electric field increases, the electrostatic repulsion on the surface of the droplet increases. When the electrostatic repulsion is stronger than the surface tension, the droplet will deform into a cone and form jets. After that, Taylor published two articles to preliminarily explore the behavior of the jet in the electric field in 1966 and 1969 [26,27].

The electrospinning method is the polymer-eruption electrostatic drawing spinning method. Firstly, the polymer solution or melt is subjected to a high-voltage electrostatic charge ranging from several thousand to ten thousand volts, and the charged polymer droplets are accelerated at the Taylor cone pole of the capillary under the effect of an electric field force. When the electric field force is large enough, the polymer droplets overcome the surface tension to form an eruptive trickle. The thin stream evaporates or solidifies in the process of eruption and eventually falls on the receiving device to form a nonwoven-like fiber felt. The dynamics of an electrically charged jet in an electric field are complex. When the jet is first formed, it is stretched along the electric field. At this segment, the jet can keep moving in a straight line. But the jet moves radially outward at a comparable velocity because of the electrical bending instability (Figure 2A). Around the 2000s, many studies attempted to build a physical or mathematical model of the jet at this segment [17,18,28–32]. The linear jet will bend under perturbation. The charge carried with the bend segment moves downward and outward by the repulsion (F_{DO}) above the bend region, and at the same time, the charge moves upward and outward by the repulsion (F_{UO}) below the bend region. The result of the two forces FR is radial and increases exponentially with the increasing degree of jet bending (Figure 2B) [16]. As a result, the jet appears as a spiral with a cone envelope. When the jet continues to move in the electric field, the jet that has been bent into a spiral shape will obtain the second bending instability, as indicated at the end of the jet in Figure 2A. The jet is stretched thinner and thinner in the process.

There are many factors affecting the preparation of nanofibers by electrospinning, which can be divided into solution properties (such as viscosity, elasticity, electrical conductivity, and surface tension), control variables (such as the static voltage in the capillary, the potential of the capillary port, and the distance between the capillary port and the collector), and environmental parameters (such as solution temperature, air humidity and temperature in the spinning environment, and airflow speed). The main influencing factors include: (1) The concentration of polymer solution: The higher the concentration of polymer solution, the greater the viscosity and the greater the external tension. The droplet cleavage can be weakened by increasing the external tension after leaving the nozzle. Generally, when other conditions remain constant, the fiber diameter increases with the increase in concentration; (2) Electric field strength: With an increase in electric field strength, the jet of the polymer electrospinning solution has a larger outer charge density and thus has a greater electrostatic repulsion. At the same time, the higher electric field strength makes the jet obtain a greater acceleration degree. Both of these factors can cause the jet and fiber to have greater tensile stress, resulting in a higher tensile strain rate, which is conducive to the preparation of finer fibers; (3) The interval between the capillary port and the collector: After the polymer droplets are ejected through the capillary port, they volatilize in the air with the solvent, and the polymer is concentrated and solidified into fibers, which are finally accepted by the receiver. As the distance between the two increases, the diameter decreases; (4) Activity rate of electrospinning fluid: When the diameter of the spinneret is fixed, the average jet velocity is obviously proportional to the diameter of the fiber; (5) The condition of the collector: The condition of the collector is different, and the condition of the produced nanofibers is also different. When the fixed collector is used, the nanofibers appear to have random irregular scenes. When using a rotating disk collector, the nanofibers appear in a parallel pattern. Therefore, the fibromomentum produced by different equipment conditions is different.

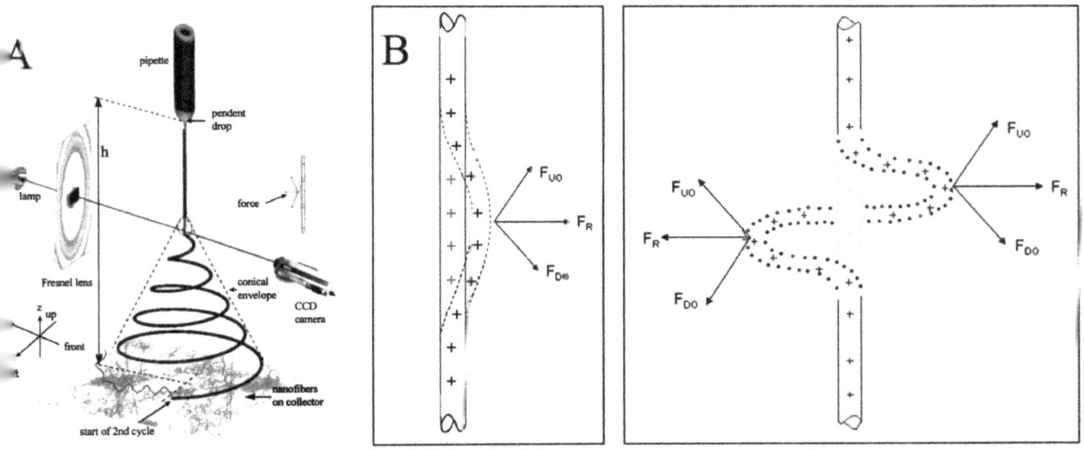

Figure 2. (**A**) An electrospinning jet that contained two successive electrical bending instabilities; (**B**) affected by the repulsion forces between charges, the perturbation segment of the jet (the dashed segment) is affected by the repulsion forces F_{UO} and F_{DO} of the lower and upper charges. The resultant of these forces FR is in the radial direction of the jet, which makes the jet more curved. (**A**) Reprinted with permission from [17]. Copyright 2000, American Institute of Physics. (**B**) Reprinted with permission from [16]. Copyright 2008, Elsevier.

In addition to spiraling, the jet will also undergo other shape changes, such as branching and the formation of beads. The undulations increase as the charge density of the jet increases. When the undulations are large enough to become unstable, the jet will branch off. The undulations come from the combined effects of the electric Maxwell stresses and surface tension [31]. Branching occurs more frequently in viscous solutions and at high electric fields. In contrast, when the charge density of the jet is reduced, surface tension will dominate in the competitive relationship between the electric field and surface tension. Capillary instability that results in the transformation of the jet into spherical droplets occurs.

In recent years, researchers have tried to predict the diameter of electrospun nanofibers in various ways, including theoretically or experimentally. Gadkari summarized the literature on various correlations and analyzed them to obtain the relationship dependency of the nanofiber diameter on the viscous and surface charge repulsion effects [33]. A design of the experiments model, analyzing data based on polynomial equations without physical meaning, was proposed by Ruiter to identify the effects of electrospinning parameters on scaffold morphologies of poly-D,L-lactic acid [34]. The model shows that solution concentration plays an important role in the morphology of nanofibers.

3. Regulation of Nanofibers

Diameter, as the most basic parameter of nanofibers, has always been the object of greatest concern to researchers. A variety of investigations have been reported on the effect of electrospinning parameters on the diameter of the nanofibers, both theoretically and experimentally. The diameter of the nanofibers is influenced by the concentration of polymer in solution, the type of solvent, the conductivity, as well as the feeding rate of the solution [35,36]. The higher the concentration and viscosity of the solution, the higher the voltage required to overcome its own surface tension to form the jet; the voltage needs to be optimized to match with the feeding rate of spinning fluid and the collecting distance for producing nanofibers with a uniform diameter.

3.1. Diameter of Nanofibers

It has been reported that the larger the voltage, the smaller the diameter of the nanofibers [37]. However, there was also the observation that the diameter of the nanofibers increased with an increase in the voltage [38]. The reason is that the type of spinning solution and the parameters are not the same. This indicates that the influence of the voltage summarized by simple control variables on the final diameter of the nanofibers is not rigorous enough. Gu et al. reported that the concentration of the spinning fluid has a greater effect on the diameter of the nanofibers than the voltage [39]. In fact, the main factor affecting the diameter of the nanofibers here is the viscosity of the solution, which changes with the concentration of the solution and the molecular weight of the polymer [39,40].

Fridrikh et al. reported that there is a limiting diameter for the jet, which arises from a force balance between surface tension and electrostatic charge repulsion [41]. These two forces are determined by the viscosity and conductivity of the solution, respectively. In the case of ignoring the elastic effect and fluid evaporation, the diameter of the nanofibers can be given by the following equation:

$$h_t = \left(\gamma\bar{\varepsilon}\frac{Q^2}{I^2}\frac{2}{\pi(2\ln\chi - 3)}\right)^{1/3} \quad (1)$$

where γ is surface tension of the solution, $\bar{\varepsilon}$ is the dielectric constant, Q is the flow rate of the solution, I is the current carried by the jet, and χ is proportional to the ratio of the jet diameter h and radius of curvature R. It is shown that the nanofiber diameter decreases with the increase in the charge carried by the jet. However, the electrospinning process is much more complex than this model, so this model is not accurate or comprehensive enough. For example, temperature is an easily overlooked parameter. Most solution electrospinning processes take place at ambient temperature. As the temperature increases, the viscosity and surface tension of solution decrease, leading to a reduction in nanofiber diameter. But at the same time, the increase in temperature will also cause the acceleration of the evaporation of the solvent. In this case, the drawing process, which is also the fining process of the nanofibers, will end prematurely. A temperature equilibrium that minimizes the diameter of the nanofibers can be found [42].

3.2. Morphologies and Structures of Electrospun Nanofibers

Electrospinning has become one of the main ways to prepare nanofibers because of its advantages such as a simple manufacturing device, good adjustable fiber structure, and strong technical combination. It can be used for directly and continuously spinning polyester, polyurethane, polyethylene, and other polymers into ultra-fine and uniform fibers with diameters ranging from less than 3 nm to more than 1 μm and depositing on the receiving plate to obtain non-woven fabric. It is also possible to prepare nanofibers with special shapes by electrospinning in addition to nanofibers with uniform diameters. Porous nanofibers, hollow nanofibers, and bead-on-string nanofibers are produced to meet the requirements of various applications. More complex structures tend to yield better performance, such as a greater specific surface area, rougher surface, or greater surface energy. Next, we will discuss the strategies for producing nanofibers with different morphologies and structures.

3.2.1. Bead-on-String Nanofibers

By adjusting the parameters in the process of electrospinning, electrospun nanofibers with unique morphologies can be simply obtained. Bead-on-string nanofibers are typical. As we discussed earlier, the jets need to overcome capillary instability to form uniform nanofibers. Otherwise, the bead structure will be obtained. At first, the bead structure appears randomly, which is considered an undesirable by-product. It is not until the production of uniform bead-on-string structures that attention is paid to this structure (Figure 3A). Zuo et al. reported how and why this structure was formed [43]. As indicated in Figure 3B, the jet is photographed at different distances from the nozzle. As the jet flows further and further away from the nozzle, the smooth surface of the jet flow first becomes wavy, then dumbbell-like, and finally bead-shaped.

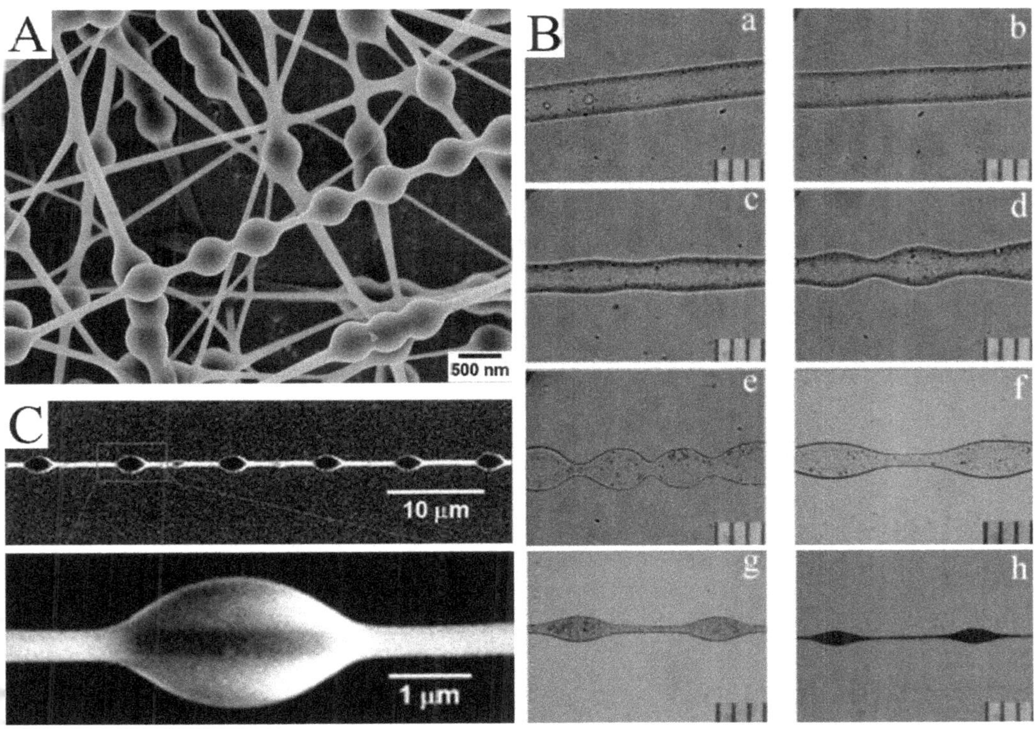

Figure 3. (**A**) SEM images of the bead-on-string nanofibers with polyvinyl alcohol (PVA): polystyrene (PS) (1:1) and 9 wt.% PVA for 473 nm PS nanospheres. (**B**) The bead formation process at different points along the PHBV nanofibers formed by electrospinning. The photos were taken at different distances from the needle tip according to their orders: (**a**) 1 cm; (**b**) 3 cm; (**c**) 5 cm; (**d**) 7 cm; (**e**) 9 cm; (**f**) 12 cm; (**g**) 15 cm; (**h**) 30 cm. (**C**) SEM image of PS nanofibers with a regularly distributed polyethylene

glycol (PEG) droplet on it. (**A**) Reprinted with permission from [44]. Copyright 2012, American Chemical Society. (**B**) Reprinted with permission from [43]. Copyright 2005, Society of Plastics Engineers. (**C**) Reprinted with permission from [45]. Copyright 2011, WILEY-VCH.

The bead-on-string structure can be regulated by controlling viscosity, net charge density, and surface tension [46]. Charge repulsion and electric field force are the causes of the drawing and deformation of the jet. Surface tension always makes the liquid the smallest surface area by turning into a sphere, while viscoelastic force resists rapid changes in shape. Therefore, the easiest strategy for producing bead-on-string nanofibers is adjusting the solution concentration and applying voltage. A higher concentration enhances electrostatic repulsion forces and viscoelasticity but reduces surface tension. A polymer solution with a lower concentration tends to produce beads with a higher density [47]. The effect of the viscosity of spinning solution on bead structure can be clearly reflected in an example. By combining electrospinning and electrospraying, Tian et al. obtained PEG beads on PS-string hetero-structured nanofibers in a coaxial jetting process (Figure 3C) [45]. During electrospinning, the outer PEG with a low viscosity is deformed by humidity, while the PS with a high viscosity keeps the shape of a "string".

The surface tension and viscoelasticity of the jet can be changed by adjusting the proportion of solvent. For example, PS-b-poly(ethylene butylene)-b-PS triblock copolymer solutions, which can be dissolved in different proportions of tetrahydrofuran/N,N-dimethylformamide (THF/DMF), have different rheology data. Accordingly, the viscosity of the solution will vary considerably [48]. In addition, a corona discharge can be used to add neutralizing charge to the jet so as to reduce electrostatic repulsion [46].

3.2.2. Porous Nanofibers

In certain cases, the rapid evaporation of the solvent causes the liquidation of water vapor during the spinning process. Droplets formed by water vapor liquefaction adhere to the jet surface, leading to the formation of pores on the fiber's surface (Figure 4A) [49]. This approach, which is essentially vapor-induced phase separation, has to meet two requirements: humid environment and volatile solvents, such as THF, methylene chloride, and carbon disulfide. The size and number of pores can be easily controlled by changing the humidity. The higher the humidity, the larger the pore size and the higher the pore density [50]. In addition to controlling environmental humidity, there are other methods of inducing phase separation. McCann et al. electrospun PAN nanofibers into liquid nitrogen [51]. In this process, pores were introduced into the nanofibers surface by thermally induced phase separation. Pores could not only be formed on the surface of the nanofibers but also inside the nanofibers so as to obtain nanofibers with a loose structure. As shown in Figure 4B, Pai et al. used PS/DMF solution to produce fibers with a smooth surface and a porous interior [52]. Compared with a highly volatile solvent mentioned earlier, the evaporation rate of DMF is not fast enough to form small droplets on the jet surface. DMF absorbed water vapor in the air due to a non-solvent phase separation reaction between them. As a result, a porous structure was formed inside the fibers. The growth process of porous structures resulting from phase separation was calculated by Dayal et al. based on the Cahn–Hilliard time-evolution equation and solvent evaporation rate equation [53]. The strategy of introducing pores through phase separation has been applied to a variety of polymers, such as PS [54], poly(L-lactic acid) [55,56], polyvinyl butyral [57], polyethylene terephthalate [58], and poly(ε-caprolactone) [59,60].

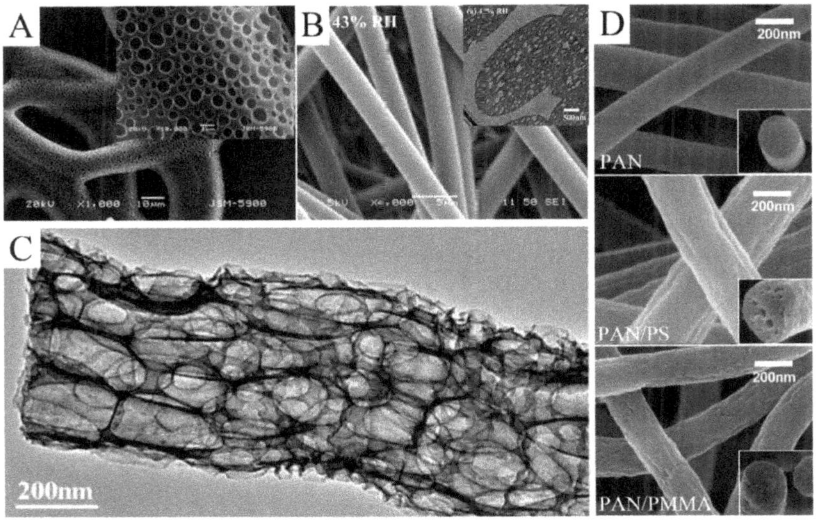

Figure 4. (**A**) SEM images of electrospun poly(ε-caprolactone) fibers collected in water as a function of pH 3. (**B**) SEM images of as-spun fibers electrospun from a 30 wt.% PS/DMF solution under 43% relative humidity. The illustration is the cross-sectional TEM images of them. (**C**) TEM images of the macroporous CNFs obtained by etching SiO_2/Sb@CNF composites with an aqueous HF solution. (**D**) SEM images of the porous CNFs produced from pure PAN, PAN/PS, and PAN/PMMA, respectively. (**A**) Reprinted with permission from [61]. Copyright 2011, Springer. (**B**) Reprinted with permission from [52]. Copyright 2009, American Chemical Society. (**C**) Reprinted with permission from [62]. Copyright 2018, American Chemical Society. (**D**) Reprinted with permission from [63]. Copyright 2016, Elsevier.

Porous nanofibers can also be produced through the removal of a sacrificial phase. Commonly, the sacrificial phases are salt, polymers, and nanoparticles. A sacrificial phase of salt can be removed by leaching. Gupta et al. used $GaCl_3$ as a template to prepare porous nylon-6 fibers [64]. Wang et al. prepared SiO_2/Sb@CNFs by carbonizing electrospun precursors, which were etched with HF solution to remove SiO_2 and Sb nanoparticles [62]. What was left were porous CNFs (Figure 4C). The porous CNFs could also be prepared by electrospinning PAN blending with the sacrificial PS or PMMA [63]. PS and PMMA were easier to pyrolyze than PAN. As a result, PS or PAMM was completely pyrolyzed to form pores during carbonization. Moreover, PS foam (Figure 4D) and Nafion are also suitable as a sacrificial phase in PAN solution [65,66].

The introduction of pores has other advantages in addition to greatly increasing the specific surface area of the nanofibers: the functionalization inside the pores is easy to realize for obtaining functional surfaces [67]; and porous CNFs have stronger ionic adsorption and higher specific capacitance [65].

3.2.3. Hollow Nanofibers

Hollow nanofibers are another common structure of electrospun nanofibers. There are two strategies for preparing hollow nanofibers. One is coaxial electrospinning, which we will discuss later. The other strategy is to make the components of the solution spontaneously move along the radial direction of the jet and be stratified by adjusting the ratio of a mixed polymer solution. Niu et al. produced hollow PVA nanofibers by mixing PVAs with different molecular weight [68]. The schematic diagram of the formation of the hollow structure is displayed in Figure 5A. With the same electrospinning parameters, the higher the molecular weight, the higher the distribution in the outer layer. The nanofibers were then carbonized at high temperatures. In this process, the low-molecular weight PVA in the inner layer first pyrolyzed and moved to the high-molecular weight PVA. Hollow CNFs were produced. It is of concern that if inorganic materials were mixed into the solution and the nanofibers carbonized in the protection of argon, the inorganic materials would not move with the low-molecular weight PVA. The inorganic materials aggregated into spheres in the channels of hollow nanofibers, and pea-like nanofibers could be formed. Hollow nanofibers are commonly prepared by similar methods. For example, the fibers have been produced by electrospinning with a mixture of polycarbosilane (PCS) and PS, with PCS in the outer layer and PS in the inner layer. After carbonization, hollow SiC nanofibers were formed, as PCS turned into SiC and PS was pyrolyzed (Figure 5B) [69]. Hollow nanofibers can also be produced based on vapor-induced phase separation. In another example, camphene and tetraethoxysilane, as pore-forming agents, could be added into PS solution. Water entered the surface of the jet in the electrospinning process. At this point, tetraethoxysilane that diffused to the outer layer evaporated, and camphene that diffused to the outer layer was removed after freeze-drying. Hollow and porous structures of electrospun PS fibrers were obtained [70].

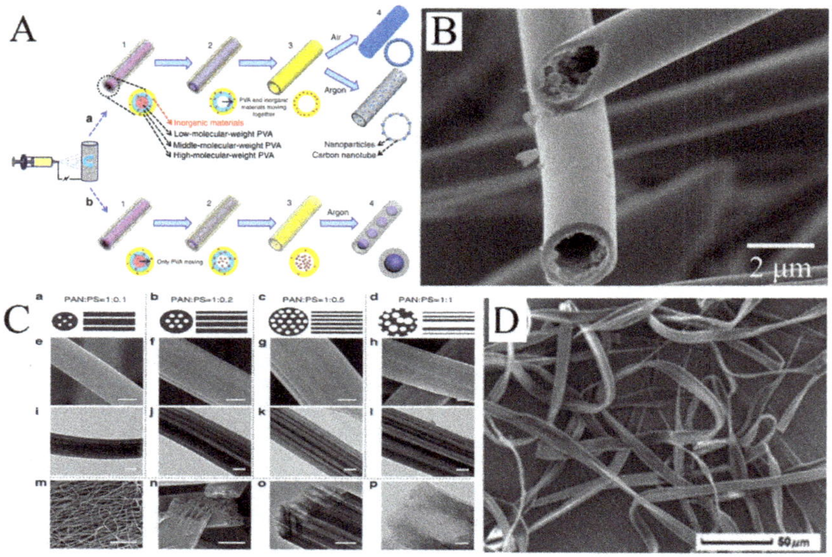

Figure 5. (**A**) Schematics of the gradient electrospinning and controlled pyrolysis method. (**B**) SEM images of hollow silicon carbide nanofibers. (**C**) (**a–d**) Schematic diagrams, (**e–h,m,n**) FESEM and (**i–l,o,p**) TEM images of LRC nanofibers based on various PAN/PS weight ratio: (**a,e,i**) 1:0.1, (**b,f,j**) 1:0.2, (**c,g,k,m–p**) 1:0.5, (**d,h,l**) 1:1. Scale bars, 200 nm (**e–l,o**), 20 μm (**m**), 500 nm (**n**), 20 nm (**p**). (**D**) SEM images of microfibers electrospun from PMAGH-block-PS-block-PMAGH in 0.20 g mL^{-1} $CHCl_3$ solution at 20 μL min^{-1} feeding rate. (**A**) Reprinted with permission from [68]. Copyright 2015, Macmillan. (**B**) Reprinted with permission from [69]. Copyright 2018, Elsevier. (**C**) Reprinted with permission from [71]. Copyright 2015, WILEY-VCH. (**D**) Reprinted with permission from [72]. Copyright 2019, WILEY-VCH.

Some hollow fibers with more interesting structures have been reported by electrospinning. For example, hollow fibers with multiple channels produced with a mixture of PAN and PS have been reported. PS as the sacrificial phase will form more than one channel inside PAN. The number and distribution of channels will vary with the ratio of PAN to PS in the solution (Figure 5C) [71]. In addition, it is possible to obtain ribbon fibers in the production of hollow fibers, and unique polymer skin can form on the surface of the jet. After that, the evaporation of solvent will result in the vacuum inside, leading to the collapse of skin (Figure 5D) [72–74].

4. Application of Electrospun Nanofibers in Electrochemical Biosensors

Electrochemical biosensors have attracted more and more attention due to their speed, simple operation, and low cost. Electrospun nanofibers have been extensively applied in electrochemical biosensors for the determination of various molecules such as glucose, hydrogen peroxide (H_2O_2), uric acid (UA), dopamine (DA), ascorbic acid (AA), protein, and amino acids.

4.1. Glucose Sensors

As a source of energy in the body, glucose is widely found in every corner of our life. People who want to lose weight need to know the amount of glucose in their diet in order to control their diet properly. Diabetics need to regularly measure their blood sugar levels to keep themselves healthy.

Glucose oxidase (GOx) is one of the main choices for real-time glucose monitoring owing to its high glucose selectivity. Nanofibers, due to their high specific surface area and porous structure, are excellent platforms for enzyme immobilization. The activity of the immobilized enzyme greatly affects the performance of the sensor. As implied in Figure 6A, the activity of GOx immobilized on the surface of PVA/malonic acid nanofibers with different plasma processing was quite different [75]. In addition, the reusability and storage stability of the plasma-treated fibers was significantly improved (Figure 6B,C). When designing the glucose sensor based on GOx, the nanofibers as the substrate have great influence on the performance of the sensor. Guo et al. produced CNFs with uniformly embedded TiC nanoparticles (Figure 6D,E) [76]. The robust adhesion of the composite nanofibers provides more abundant active sites for enzyme immobilization (Figure 6F). The GOx-TiC-CNFs biosensor could selectively detect glucose with a wide linear range (0.013–10.5 mM) and low detection limit (3.7 μM).

Enzymes are unstable at certain temperatures and pH values. Therefore, non-enzyme sensors have been developed in which enzymes are generally replaced by metals or metal oxides. Non-enzyme sensors are more economical than enzymatic sensors and are less affected by the environment. Glucose is the most popular analyte. A variety of transition metals are used in non-enzyme electrochemical sensors for glucose. The mechanism diagram of the transition metal electrocatalytic detection of glucose is illustrated in Figure 7A. For example, Lu et al. reported a glassy carbon electrode (GCE) modified with CuO/Cu_2O composite nanofibers for a glucose sensor [77]. CuO/Cu_2O composite nanofibers had good electrocatalytic activity for glucose. The possible electrocatalytic processes of glucose can be explained as follows: The electrons produced by the oxidation of Cu^{2+} to Cu^{3+} form an oxidation peak in the cyclic voltammograms. After that, Cu oxidizes glucose to gluconolactone, which further increases the oxidation current [78–80]. In a specific range of glucose concentration, the oxidation peak increases linearly with the increase in glucose concentration. This property enables the determination of glucose concentration by the magnitude of the oxidation current (Figure 7B) [81]. Increased conductivity can accelerate electron transfer and improve the performance of the sensors. CNFs have excellent mechanical and electrical properties and can be simply obtained using carbonized electrospun polymer nanofibers, making them suitable as carriers for metal oxide nanoparticles. In addition, the synergistic effect of multiple components can also greatly enhance the electrocatalytic performance. The reason may be that the defects caused by the lattice mismatch of the two metal atoms lead to electron accumulation at the interface and the upward shift of the D band. Shi et al. prepared electrospun CuO nanofibers, CuO/CNFs, and CuO/NiO nanofibers as a glucose sensor, as shown in Figure 7C [82]. The electrocatalytic effect can be observed through the magnitude of the catalytic current and the position of the oxidation peak potential. Obviously, CuO/NiO nanofibers can produce higher current at lower potential.

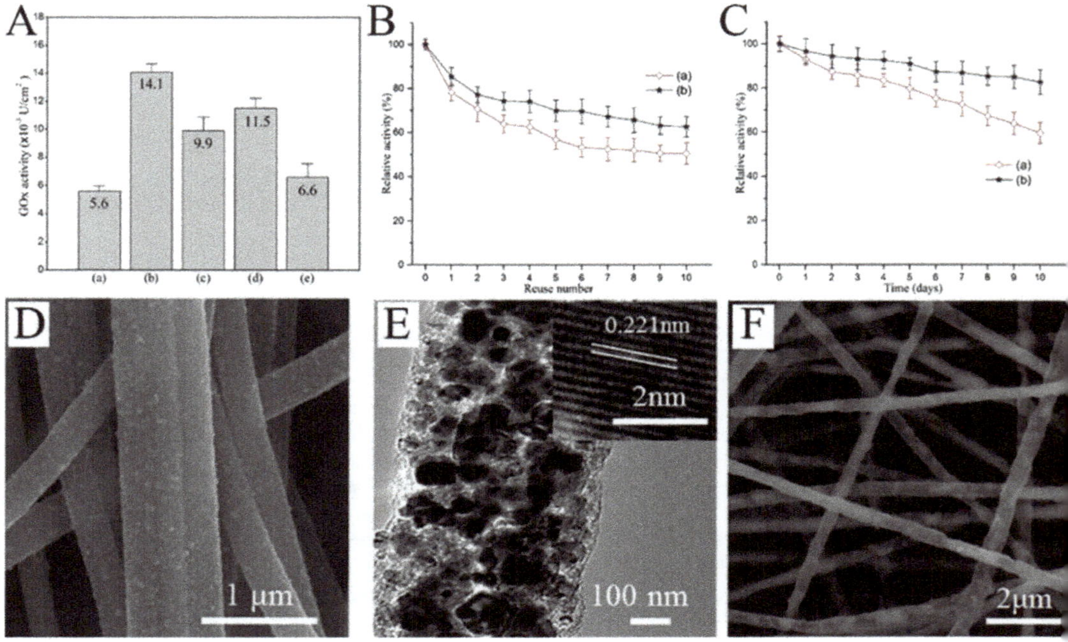

Figure 6. (**A**) Activity of GOx immobilized on the surfaces of (**a**) untreated and (**b**) air-, (**c**) nitrogen-, (**d**) CO_2-, and (**e**) argon plasma-treated PVA/malonic acid nanofibers. (**B**) Reusability of GOx immobilized on (**a**) unmodified and (**b**) air plasma-modified nanofibers. (**C**) Storage stability of GOx immobilized on (**a**) unmodified and (**b**) air plasma-modified nanofibers. (**D**) SEM images of TiC CNFs. (**E**) TEM images of TiC CNFs; inset of (**E**) is the HRTEM image. (**F**) SEM image of GOx–TiC CNFs. (**A–C**) Reprinted with permission from [75]. Copyright 2016, Elsevier. (**D–F**) Reprinted with permission from [76]. Copyright 2018, Elsevier.

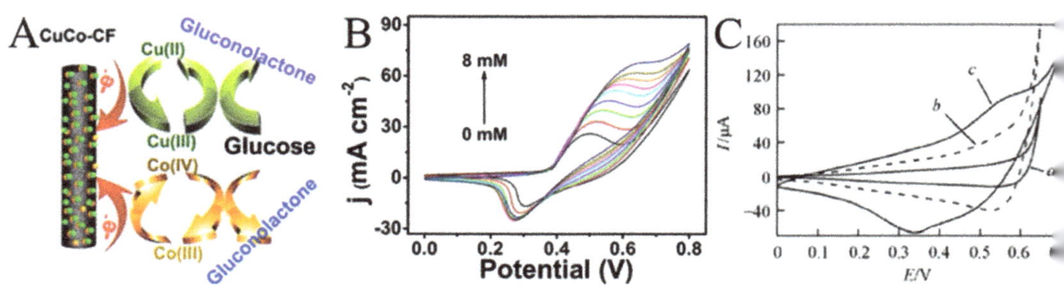

Figure 7. (**A**) Schematic diagram of the transition metal electrocatalysis detection of glucose. (**B**) CVs for Ni_2P nanofibers in 0.1 M NaOH (pH 13) with the presence of varying glucose concentrations ranging from 0 to 8 mM. (**C**) CVs of (a) CuO-NFs, (b) CuO/CNFs, and (c) CuO/NiO-NFs film electrodes in 0.1 mol L^{-1} NaOH containing 0.6 mmol L^{-1} glucose at 50 mV s^{-1}. (**A**) Reprinted with permission from [80]. Copyright 2016, American Chemical Society. (**B**) Reprinted with permission from [81]. Copyright 2016, American Chemical Society. (**C**) Reprinted with permission from [82]. Copyright 2013, WILEY-VCH.

There have been many reports on the application of electrospinning to non-enzyme electrocatalysts for the detection of glucose. Saravanan et al. prepared Co-Fe/PVdF-HFP by electrospinning and chemical reduction techniques for electrochemical glucose detection, in which the diffusion and adsorption of glucose in the expanded cavities and pores of the polymer nanofibers were accelerated to maximize the utilization efficiency of glucose [83]. Luo et al. fabricated a Pt-Au/polyurethane sensing patch through electrospinning, magnetron sputtering, and electrodeposition techniques for the determination of glucose in a neutral condition (pH 7.4) [84]. Li et al. synthesized hollow CuO/NiO nanoparticles with adjustable sizes by coaxial electrospinning and subsequent calcination [85]. The unique morphology and high specific surface area as well as the hetero-structural interface between CuO and NiO are conducive to improving the electrocatalytic performance, showing excellent electrocatalytic performance for glucose oxidation. They also prepared nano-Mn_3O_4/NiO-decorated CNFs by electrospinning and calcination [86]. The conductive network constructed by CNFs not only promotes the transfer of electrons but also provides a landing site for nanoparticles, thereby reducing the aggregation of nanoparticles and exposing more active sites. Kim et al. embedded MnO nanostructures with CNFs to significantly improve the detection performance of non-enzymatic amperometric glucose sensors [87]. Additionally, other nanofibers are also used in glucose biosensors, such as Ni_2P/CNFs [88], PAN/PANI/CuO [89], CuSn/CNFs [90], and CuCo-P350 [91]. Table 1 summarizes several examples of non-enzyme electrochemical glucose sensors, including the detection limit, sensitivity, linear range, and oxidation potential.

Table 1. Electrospun nanofibers for non-enzyme glucose electrochemical biosensors.

Catalyst	LOD (μM)	Sensitivity (μA mM^{-1} cm^{-2})	Linear Range (μM)	Potential (V)	Ref.
Co-Fe/PVdF-HFP	0.65	375.01	1–8000	0.53	[83]
TiO_2/Cu_2O/CuO CNFs	0.25	2074.7	0–2000	0.55	[80]
Pt-Au/polyurethane	14.77	203.13	0.1–50	N/R	[84]
CuO/NiO NFs	14.77	1324.17	1–1000	0.6	[85]
Mn_3O_4/NiO/CNFs	0.73	386 84	5–3000	0.5	[86]
MnOx-CNFs	0.3	4080.6	0–9100	0.55	[87]
Ni_2P/CNFs	0.25	1050	5–208	0.5	[88]
PAN/PANI/CuO	1.2	N/R	3–500	0.4	[89]
CuSn/CNFs	0.08	N/R	0.1–9000	0.55	[90]
CuCo-P350	2.92	2272	5–825	0.55	[91]

4.2. H_2O_2 Sensors

In organisms, H_2O_2 is a signaling molecule that controls cell metabolism. In addition, it is also associated with some human diseases, such as cardiovascular diseases, Alzheimer's Disease, and cancers [92]. Therefore, the detection of H_2O_2 is of great significance in medical applications.

Noble metal (such as Au [93], Ag [94], Pd [95], and Pt [96]) nanoparticles have excellent electrocatalytic activity with H_2O_2. For example, Huang et al. electrospun Pd nanoparticle-loaded CNFs to modify a carbon paste electrode (CPE) [97]. The modified CPE shows a wide linear range (0.2 μM to 20 mM) for H_2O_2 (Figure 8A). As shown in Figure 8B, Zhang et al. compared the cyclic voltammetry of H_2O_2 on PVDF, a multi-walled carbon nanotube (MWCNT), and Pt [98]. It was not difficult to see that the electrocatalytic performance of H_2O_2 mainly came from Pt. Without Pt, PVDF or MWCNT alone showed almost no current response to H_2O_2.

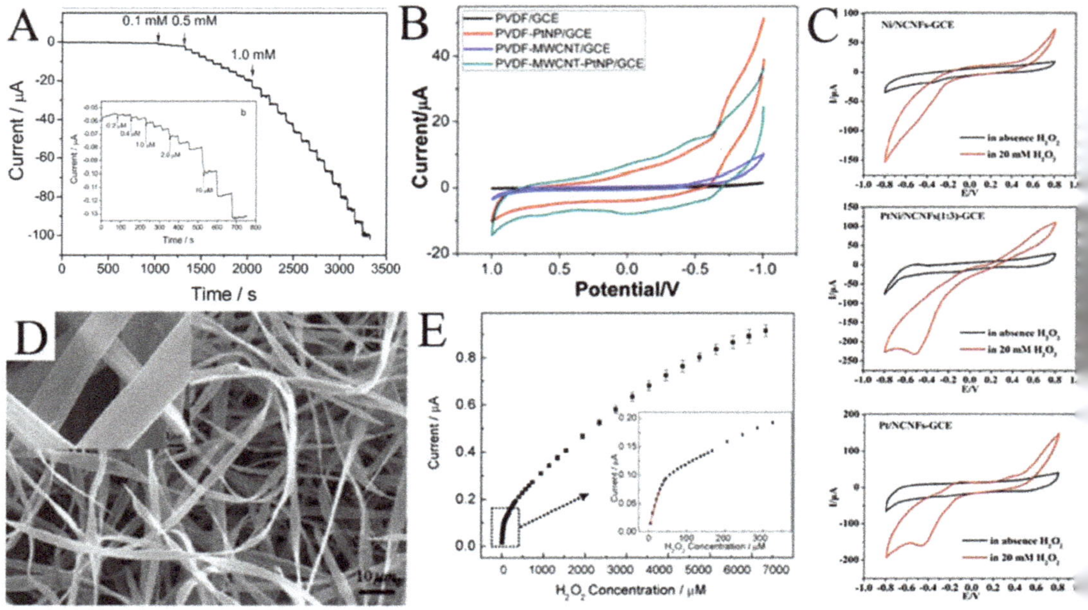

Figure 8. (**A**) Current–time responses of the Pd/CNFs-CPE upon successive injection of a specific concentration of H_2O_2 into N_2-saturated PBS (0.1 M, pH 7.0); the inset (**b**) shows the performance of the Pd/CNFs-CPE in the amperometric detection of a low concentration of H_2O_2. Applied potential: −0.2 V. (**B**) CVs of GCEs modified with PVDF, PVDF-PtNP, PVDF-MWCNT, and PVDF-MWCNT-Pt hybrid nanofiber membranes. (**C**) CVs of Ni/CNFs-GCE, PtNi/CNFs-GCE, and Pt/CNFs-GCE in the absence and presence of H_2O_2 at 50 mV s^{-1}. (**D**) Typical SEM images of Hb microbelts. (**E**) The corresponding calibration plot of amperometric response towards H_2O_2. Inset: enlarged drawing of the calibration plot for low H_2O_2 concentrations. (**A**) Reprinted with permission from [97]. Copyright 2008, WILEY-VCH. (**B**) Reprinted with permission from [98]. Copyright 2014, American Chemical Society. (**C**) Reprinted with permission from [99]. Copyright 2018, Elsevier. (**D**,**E**) Reprinted with permission from [100]. Copyright 2010, Elsevier.

High cost limits the development of H_2O_2 sensors based on noble metals, so various transition metals/metal oxides are introduced to functionalize noble metals. As shown in Figure 8C, compared with Pt/CNFs and Ni/CNFs, the electrocatalytic performance of PtNi/CNFs for H_2O_2 is much better [99]. The PtNi/CNFs-modified GCE has a wide detection range from 0.05 μM to 8 mM with a low detection limit of 0.0375 μM for H_2O_2. Mohammadi et al. synthesized a Se/P@N-doped carbon nanobox (N-CNB)/CNFs for the electrochemical determination of H_2O_2 [21]. Due to the mesoporous structure, the open efficient diffusion channel for H_2O_2, the rapid mass/electron transfer, and the synergies between P, Se, and CNBs/CNFs, the synthesized Se/P@(N-CNB)/CNFs exhibited excellent electrocatalytic activity toward H_2O_2 oxidation. Hsueh et al. prepared IrO_2@Ir NFs for the selective determination of H_2O_2 [86]. Qu et al. synthesized LaSrNiO NFs with a dual-phase structure and unique porous tubular nanofiber structure using electrospinning and high-temperature calcination techniques, significantly improving their redox performance for the electrochemical sensing of H_2O_2 [101]. Bi et al. modified the conductive polymer PEDOT:PSSLiTFSI-CoPc (PPLC) on nanofibers (PPLC/PU/PDMS) to develop a stretchable electrochemical sensor [102]. The electrode displayed good electrochemical sensing performance and stability under mechanical deformation. In addition, Ag@CuO [103], VCoO/C-750 [104], and Co-NC/CNF [105] are also used for H_2O_2 detection. Table 2 lists several examples of H_2O_2 sensors using electrospun metals/metal oxides nanofibers.

Table 2. Electrospun nanofibers for H_2O_2 analysis.

Catalyst	LOD (μM)	Sensitivity (μA mM^{-1} cm^{-2})	Linear Range (μM)	Potential (V)	Ref.
Se/P@CNBs/CNFs	58	171.1	200–1800	NR	[21]
IrO_2@Ir NFs	0.16	289	0.1–1000	−0.4	[86]
PtNi/CNFs	0.0375	248.5	0.05–8000	−0.1	[99]
LaSrNiO NFs	0.018	1667.9	1–7000	0.2	[101]
PPLC/PU/PDMS	0.2	0.0406	0.5–50	+0.55	[102]
Ag@CuO	0.01	1982.14	0.05–100	+0.6	[103]
VCoO/C-750	0.44	N/R	0–8000	N/R	[104]
Co-NC/CNF	10	300	1–6000	−0.5	[105]

Interestingly, hemoglobin microbelts have been successfully electrospun for H_2O_2 sensing by Ding et al. (Figure 8D) [100]. In their work, hemoglobin is dissolved in 2,2,2-trifluoroethanol as the spinning solution for electrospinning hemoglobin microbelts, allowing for the sensitive detection of H_2O_2 at physiological pH (0.1 M pH 7.0 phosphate buffer). As shown in Figure 8E, this H_2O_2 sensor has a low detection limit (0.61 μM) and high stability due to its good biocompatibility and direct electron transfer capability.

4.3. Detection of Other Biomolecules

A variety of electrospun nanofibers have been developed for the detection of biomolecules, such as UA, DA, AA, protein, and amino acids. The level of these biomolecules in the body can reflect the health of the body and help doctors determine whether the body is suffering from certain diseases. For example, the concentration level of UA is one essential indicator for the diagnosis and the prognosis of some multifunctional disorders like gout, hypertension, and cardiovascular diseases [106]. DA is an important neurotransmitter, having strong influence on the central nervous, renal, cardiovascular, and hormonal systems [107].

CNFs have been widely used in electroanalysis due to their good dispersibility, wettability, conductivity, and biocompatibility. In some cases, the electrochemical performance of CNFs may be even better than that of CNTs (Figure 9A) [107,108]. CNFs can be easily obtained by carbonizing electrospun polymer nanofibers at high temperature. Tang et al. prepared electrospun CNFs-modified CPE to detect trace amounts of L-tryptophan (Trp), L-tyrosine (Tyr), and L-cysteine (Cys) [109]. The modified electrode showed a satisfactory linear range (0.1–118.5 μM for Trp, 0.2–109 μM for Tyr, and 0.15–63.8 μM for Cys). As implied in Figure 9B, in addition to the obvious current response to Trp, Tyr, and Cys, the CNFs-modified electrode also showed a response to UA and AA. The results show that CNFs can detect multiple biomolecules simultaneously by DPV. Recently, effort has been devoted to improving the electrocatalytic performance of CNFs, including the combination with MWCNTs and precious metals (such as Pd and Ag) [106,110,111].

An interesting example is the dopamine biosensor based on the electrospun PANi/carbon quantum dots (CQDs) composite nanofibers [112]. On one hand, CQDs have electrocatalytic activity with dopamine. On the other hand, the addition of dopamine can cause the fluorescence quenching of CQDs. Therefore, the electrospun PANi/CQDs composite nanofibers are capable of both electrochemical and fluorescent sensors for DA (Figure 9C). In another work, Samie et al. combined a novel RuO_2-CeO_2-AuNFs hybrid structure with graphite oxide and functionalized multiwalled carbon nanotubes for the simultaneous determination of serotonin, DA, and AA [113]. The proposed electrochemical sensor reduces overpotential and solves the problem of overlapping the oxidation peak potential. Yin et al. decorated nitrogen-doped electrospun CNFs with tightly packed Co_3O_4 nanoparticles with a high electroactive surface area and promoted electron transfer between the electrode surface and the target molecule by rapid electrodeposition [114]. The electrosensing platform showed excellent sensitivity and selectivity for DA. Veeralingam et al. demonstrated a high-performance field-effect transistor biosensor based on Al-functionalized β-Bi_2O_3 nanofibers for the highly selective and rapid detection of serotonin [115]. The biotransistor has good sensitivity, stability, and repeatability and a fast response time of 0.8 s.

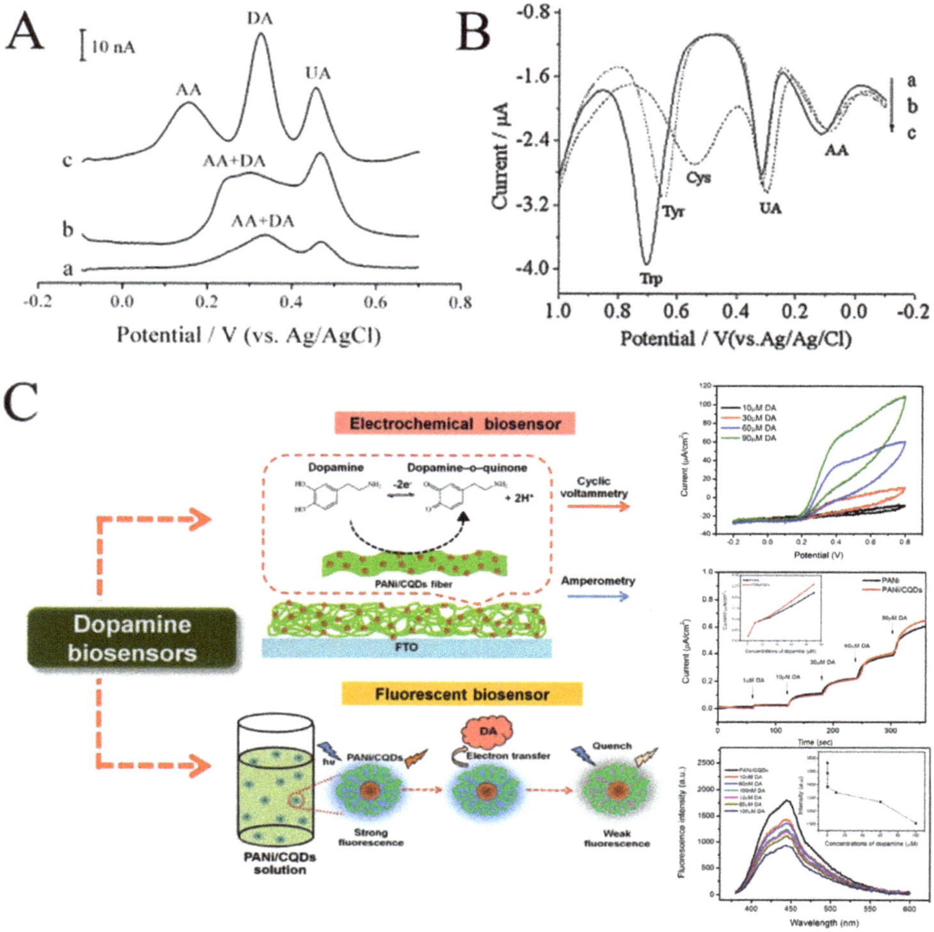

Figure 9. (**A**) DPVs at (a) CPE, (b) MWCNT-CPE, and (c) CNFs-CPE in 0.1 M PBS (pH 4.5) containing 2 μM DA, 6 μM AA, and 2 μM UA. DPV conditions: scan rate, 20 mV s^{-1}; amplitude, 50 mV; pulse width, 100 ms; pulse period, 200 ms. (**B**) DPVs of CNFs-CPE in 0.1 M PBS (pH 7.0) containing 0.1 mM UA, 0.2 mM AA, and 0.2 mM Trp (a), 1 mM Cys (b), and 0.2 mM Tyr (c). DPV conditions: Scan rate, 6 mVs^{-1}; amplitude, 50 mV; pulse width, 100 ms; pulse period, 200 ms. (**C**) Schematic diagram representing the developed electrochemical sensor and fluorescent sensor based on a PANi/CQDs composite. (**A**) Reprinted with permission from [108]. Copyright 2008, Elsevier. (**B**) Reprinted with permission from [109]. Copyright 2009, Elsevier. (**C**) Reprinted with permission from [112]. Copyright 2020, The Polymer Society.

Enzymes and antibodies could be immobilized on electrospun nanofibers for the specific identification and detection of biomolecules. For example, Mondal et al. covalently immobilized cholesterol esterase and cholesterol oxidase on electrospun TiO_2 nanofibers with a high orientation for the detection of esterified cholesterol [116]. Atilgan et al. prepared dendrimer-modified montmorillonite-decorated poly-σ-caprolactone and chitosan-based nanofibers, which were conjugated with glutamate oxidase for the electrochemical determination of monosodium glutamate [117]. Proteins can be detected by immobilizing antibodies on electrospun nanofibers through the specific binding between antigens and antibodies. For example, Macwan et al. immobilized anti-C-reactive protein on electrospun PVA/CNT nanofibers for the detection of inflammatory marker C-reactive protein [118].

5. Challenges and Opportunities Facing Practical Application

Although electrospinning technology has been deeply developed and its application has been extensively explored in many fields, there are still many issues left in both basic theory and practical application:

(i) Improvement of the theoretical model about jet: Although many models of jet movement in the electrostatic field have been established, these models are incomplete and are mostly half-empirical and half-theoretical [39,119,120]. In short, the theoretical model is still in its infancy.

(ii) Precise control of environmental factors: It has been widely reported that temperature and humidity have great influence on the morphology of electrospun nanofibers [42,121]. In the laboratory, the environmental factors of electrospinning are relatively easy to maintain. However, in order to realize the commercialization of electrospun nanofibers, large-scale production is inevitable. Under such circumstances, it is still a challenge to keep the environment stable.

(iii) Safety issues caused by solvent volatilization: A large amount of solvent volatilizes during the solidification of the jet. In large-scale production, the amount of solvent volatilization will become larger. Solvents widely used in electrospinning include DMF, DMSO, and sometimes acetone, which are toxic and may cause explosions.

(iv) Safety of nanofibers: Some research has suggested that nanofibers are likely to cause an inhalation hazard. For example, inhalation of a sufficient dose of asbestos nanofibers may cause mesothelioma [122]. At present, there are few studies on the inhalation safety of nanofibers. An in-depth study of this aspect is of great significance. Although electrospinning technology can protect our health, it can be more harmful if we ignore its own safety.

(v) Miniaturization of biosensors: With the advent of the 5G era, online medical treatment is available. People can enjoy medical services at home without the tedious procedures of offline medical treatment. Traditional testing methods usually require expensive testing equipment and complicated operations. Microscale biosensors would make online diagnosis easier. Several test papers and miniaturized instruments have been reported, but these fall far short of the need for online diagnoses [123,124]. The design of miniaturized biosensors may be a hotspot in the future.

(vi) The bionic device: Electrospun nanofiber mats mostly have excellent flexibility, which makes electrospinning technology promising in biomimetics. For example, PVDF, a piezoelectric material usually used for electrospinning, has been reported for the design of electronic skin (Figure 10A) [125–127]. There have also been a number of reports about electronic tongues, although they were only in the initial stage (Figure 10B) [128–130]. With the development of technology, it will be possible to integrate electronic tongues into real tongue sizes in the future. In addition, we can imagine that the PVDF nanofibers with piezoelectric properties could be used as not only electronic skin, but also electronic ears, an electronic throat, and even an electronic heart. On the software side, AI is developing rapidly. In this sense, bionic humans, the stuff of science fiction, may become a reality in the distant or even near future.

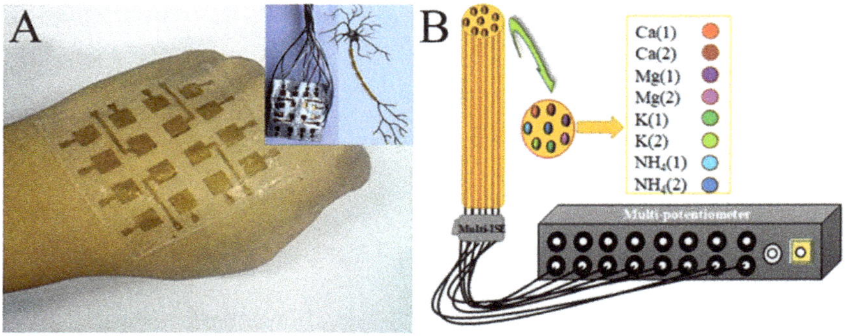

Figure 10. (**A**) Photograph of the e-skin based on a piezoelectric nanogenerator by electrospun PVDF nanofibers. The inset shows the demonstration of its analog to a neuron cell. (**B**) Schematic illustration of the multi-ion-selective electrode. (**A**) Reprinted with permission from [120]. Copyright 2018, American Chemical Society. (**B**) Reprinted with permission from [124]. Copyright 2020, Elsevier.

6. Conclusions

Electrospinning is a simple and efficient method for producing nanofibers. Electrospun nanofibers have large specific surface area, high porosity, and diversified structure. In addition, it is easy to functionalize the surface of electrospun nanofibers. These advantages make electrospinning technology have great potential in electrochemical sensors. In the past decades, a large number of studies have laid the foundation for the control of the structure and morphology of electrospun nanofibers, and many interesting composite nanofibers structures have been witnessed in practical applications of electrochemical sensing.

In this review, an overview was provided on the morphological and structural regulation of nanofibers by electrospinning technology as well as its application in electrochemical sensors. The factors affecting the morphology of the nanofibers and the strategies for producing nanofibers with special morphology were discussed. Generally, the performance of nanofibers can be improved and their application can be expanded through the following strategies: (i) nanofiber diameter; (ii) special morphology (hollow, core-sheath, porous and bead-like); (iii) functionalization of surfaces; (iv) metallization of nanofibers; and (v) carbonization to form CNFs. The working principles and advantages of various nanofibers obtained by these strategies in the detection of different substances were also discussed in this review.

In summary, many achievements have been made in both the theory and application of electrospinning technology in recent decades. Electrospinning has been involved in many fields, such as energy storage, tissue engineering, environmental protection, smart textiles, electronic devices, and physical and chemical sensors. In particular, the current research on electrospinning has a tendency of miniaturization, simplification, and economization, making it possible for the technique to play an important role in intelligent medical treatment. However, there are still many issues. For example, process and safety problems still exist in the large-scale production of electrospinning. In addition, it is necessary to establish a complete evaluation system for the safety of electrospun nanofibers.

Author Contributions: Conceptualization, J.L., K.H. and L.L.; investigation, J.L., Z.D., K.H., Z.H. and D.D; writing—original draft preparation, J.L., Z.D., K.H., Z.H. and D.D.; writing—review and editing, J.L., Z.D., D.D. and L.L.; supervision, Q.Z. and L.L.; project administration, L.L. All authors have read and agreed to the published version of the manuscript.

Funding: This research was funded by the National Natural Science Foundation of China (grant numbers 61971274 and 62171268).

Institutional Review Board Statement: Not applicable.

Informed Consent Statement: Not applicable.

Data Availability Statement: Not applicable.

Conflicts of Interest: The authors declare no conflicts of interest.

erences

Luo, C.J.; Stoyanov, S.D.; Stride, E.; Pelan, E.; Edirisinghe, M. Electrospinning versus fibre production methods: From specifics to technological convergence. *Chem. Soc. Rev.* **2012**, *41*, 4708–4735. [CrossRef] [PubMed]

Lu, N.; Jing, X.D.; Xu, Y.; Lu, W.; Liu, K.C.; Zhang, Z.Y. Effective cascade modulation of charge-carrier kinetics in the well-designed multi-component nanofiber system for highly-efficient photocatalytic hydrogen generation. *Acta Phys. Chim. Sin.* **2023**, *39*, 2207045. [CrossRef]

Wang, C.H.; Shao, C.L.; Zhang, X.T.; Liu, Y. SnO_2 Nanostructures-TiO_2 nanofibers heterostructures: Controlled fabrication and high photocatalytic properties. *Inorg. Chem.* **2009**, *48*, 7261–7268. [CrossRef] [PubMed]

Zhang, B.; Kang, F.; Tarascon, J.M.; Kim, J.K. Recent advances in electrospun carbon nanofibers and their application in electrochemical energy storage. *Prog. Mater. Sci.* **2016**, *76*, 319–380. [CrossRef]

Jian, S.; Zhu, J.; Jiang, S.; Chen, S.; Fang, H.; Song, Y.; Duan, G.G.; Zhang, Y.F.; Hou, H.Q. Nanofibers with diameter below one nanometer from electrospinning. *RSC Adv.* **2018**, *8*, 4794–4802. [CrossRef]

Ahlawat, J.; Kumar, V.; Gopinath, P. Carica papaya loaded poly(vinyl alcohol)-gelatin nanofibrous scaffold for potential application in wound dressing. *Mater. Sci. Eng. C Mater. Biol. Appl.* **2019**, *103*, 109834. [CrossRef]

Shojaie, S.; Rostamian, M.; Samadi, A.; Alvani, M.A.S.; Khonakdar, H.A.; Goodarzi, V.; Zarrintaj, R.; Servatan, M.; Asefnejad, A.; Baheiraei, N.; et al. Electrospun electroactive nanofibers of gelatin-oligoaniline/poly(vinyl alcohol) templates for architecting of cardiac tissue with on-demand drug release. *Polym. Adv. Technol.* **2019**, *30*, 1473–1483. [CrossRef]

Patel, S.; Hota, G. Synthesis of novel surface functionalized electrospun PAN nanofibers matrix for efficient adsorption of anionic CR dye from water. *J. Environ. Chem. Eng.* **2018**, *6*, 5301–5310. [CrossRef]

Thenmozhi, S.; Krishnaveni, T.; Kadirvelu, K. Reduction of nitrocompounds in aqueous medium using electrospun MgO nanofibers. *Mater. Res. Express* **2019**, *6*, 065020. [CrossRef]

Ruiz, V.; Perez-Marquez, A.; Maudes, J.; Grande, H.J.; Murillo, N. Enhanced photostability and sensing performance of graphene quantum dots encapsulated in electrospun polyacrylonitrile nanofibrous filtering membranes. *Sens. Actuators B Chem.* **2018**, *262*, 902–912. [CrossRef]

Robbins, R.J. Phenolic acids in foods: An overview of analytical methodology. *J. Agric. Food Chem.* **2003**, *51*, 2866–2887. [CrossRef] [PubMed]

Safdarian, M.; Hashemi, P.; Ghiasvand, A. A fast and simple method for determination of beta-carotene in commercial fruit juice by cloud point extraction-cold column trapping combined with UV-Vis spectrophotometry. *Food Chem.* **2021**, *343*, 128481. [CrossRef] [PubMed]

Smith, C.A.; Want, E.J.; O'Maille, G.; Abagyan, R.; Siuzdak, G. XCMS: Processing mass spectrometry data for metabolite profiling using nonlinear peak alignment, matching, and identification. *Anal. Chem.* **2006**, *78*, 779–787. [CrossRef] [PubMed]

Grieshaber, D.; MacKenzie, R.; Voros, J.; Reimhult, E. Electrochemical biosensors: Sensor principles and architectures. *Sensors* **2008**, *8*, 1400–1458. [CrossRef] [PubMed]

Kalambate, P.K.; Rao, Z.; Dhanjai; Wu, J.; Shen, Y.; Boddula, R.; Huang, Y.H. Electrochemical (bio) sensors go green. *Biosens. Bioelectron.* **2020**, *163*, 112270. [CrossRef]

Reneker, D.H.; Yarin, A.L. Electrospinning jets and polymer nanofibers. *Polymer* **2008**, *49*, 2387–2425. [CrossRef]

Reneker, D.H.; Yarin, A.L.; Fong, H.; Koombhongse, S. Bending instability of electrically charged liquid jets of polymer solutions in electrospinning. *J. Appl. Phys.* **2000**, *87*, 4531–4547. [CrossRef]

Pan, C.T.; Chang, C.C.; Yang, Y.S.; Yen, C.K.; Kao, Y.H.; Shiue, Y.L. Development of MMG sensors using PVDF piezoelectric electrospinning for lower limb rehabilitation exoskeleton. *Sens. Actuators A Phys.* **2020**, *301*, 111708. [CrossRef]

Liu, Z.H.; Pan, C.T.; Lin, L.W.; Lai, H.W. Piezoelectric properties of PVDF/MWCNT nanofiber using near-field electrospinning. *Sens. Actuators A Phys.* **2013**, *193*, 13–24. [CrossRef]

Chauhan, D.; Gupta, P.K.; Solanki, P.R. Electrochemical immunosensor based on magnetite nanoparticles incorporated electrospun polyacrylonitrile nanofibers for vitamin-D_3 detection. *Mater. Sci. Eng. C Mater. Biol. Appl.* **2018**, *93*, 145–156. [CrossRef]

Mohammadi, F.; Roushani, M.; Hosseini, H. Selenium and phosphide-doped hollow porous N-doped carbon nanobox-based electrospun N-doped carbon nanofibers toward electrochemical sensing of hydrogen peroxide. *New J. Chem.* **2024**, *48*, 8074–8081. [CrossRef]

Al-Enizi, A.M.; Elzatahry, A.A.; Abdullah, A.M.; Vinu, A.; Iwai, H.; Al-Deyab, S.S. High electrocatalytic performance of nitrogen-doped carbon nanofiber-supported nickel oxide nanocomposite for methanol oxidation in alkaline medium. *Appl. Surf. Sci.* **2017**, *401*, 306–313. [CrossRef]

Zhang, H.; Zhang, G.H.; Li, Z.Q.; Qu, K.; Shi, H.M.; Zhang, Q.F.; Duan, H.G.; Jiang, J.H. Osiers-sprout-like heteroatom-doped carbon nanofibers as ultrastable anodes for lithium/sodium ion storage. *Nano Res.* **2018**, *11*, 3791–3801. [CrossRef]

McKee, M.G.; Layman, J.M.; Cashion, M.P.; Long, T.E. Phospholipid nonwoven electrospun membranes. *Science* **2006**, *311*, 353–355. [CrossRef] [PubMed]

25. Taylor, G. Disintegration of water drops in an electric field. *Proc. R. Soc. Lond. Ser. A Math. Phys. Sci.* **1964**, *280*, 383–
[CrossRef]
26. Taylor, G. The force exerted by an electric field on a long cylindrical conductor. *Proc. R. Soc. Lond. Ser. A Math. Phys. Sci.* 1
291, 145–158. [CrossRef]
27. Taylor, G.I.; Van Dyke, M.D. Electrically driven jets. *Proc. R. Soc. Lond. A Math. Phys. Sci.* **1969**, *313*, 453–475. [CrossRef]
28. Yarin, A.L.; Koombhongse, S.; Reneker, D.H. Bending instability in electrospinning of nanofibers. *J. Appl. Phys.* **2001**
3018–3026. [CrossRef]
29. Hohman, M.M.; Shin, M.; Rutledge, G.; Brenner, M.P. Electrospinning and electrically forced jets. II. Applications. *Phys. Fl*
2001, *13*, 2221–2236. [CrossRef]
30. Yarin, A.L.; Koombhongse, S.; Reneker, D.H. Taylor cone and jetting from liquid droplets in electrospinning of nanofibers. *J.*
Phys. **2001**, *90*, 4836–4846. [CrossRef]
31. Yarin, A.L.; Kataphinan, W.; Reneker, D.H. Branching in electrospinning of nanofibers. *J. Appl. Phys.* **2005**, *98*, 064501. [Cross
32. He, J.H.; Wan, Y.Q.; Yu, J.Y. Allometric scaling and instability in electrospinning. *Int. J. Nonlinear Sci. Numer. Simul.* **20**
243–252. [CrossRef]
33. Gadkari, S.B. Scaling analysis for electrospinning. *SpringerPlus* **2014**, *3*, 705. [CrossRef]
34. Ruiter, F.A.A.; Alexander, C.; Rose, F.; Segal, J.I. A design of experiments approach to identify the influencing parameters
determine poly-D,L-lactic acid (PDLLA) electrospun scaffold morphologies. *Biomed. Mater.* **2017**, *12*, 055009. [CrossRef]
35. Mahmoudifard, M.; Vossoughi, M. Different PES nanofibrous membrane parameters effect on the efficacy of immunoa
performance. *Polym. Adv. Technol.* **2019**, *30*, 1968–1977. [CrossRef]
36. Baji, A.; Mai, Y.W.; Wong, S.C.; Abtahi, M.; Chen, P. Electrospinning of polymer nanofibers: Effects on oriented morpho
structures and tensile properties. *Compos. Sci. Technol.* **2010**, *70*, 703–718. [CrossRef]
37. Hu, J.; Wang, X.; Ding, B.; Lin, J.; Yu, J.; Sun, G. One-step electro-spinning/netting technique for controllably prepa
polyurethane nano-fiber/net. *Macromol. Rapid Commun.* **2011**, *32*, 1729–1734. [CrossRef]
38. Demira, M.M.; Yilgorb, I.; Yilgorb, E.; Erman, B. Electrospinning of polyurethane fibers. *Polymer* **2002**, *43*, 3303–3309. [Cros
39. Gu, S.Y.; Ren, J.; Vancso, G.J. Process optimization and empirical modeling for electrospun polyacrylonitrile (PAN) nano
precursor of carbon nanofibers. *Eur. Polym. J.* **2005**, *41*, 2559–2568. [CrossRef]
40. Gupta, P.; Elkins, C.; Long, T.E.; Wilkes, G.L. Electrospinning of linear homopolymers of poly(methyl methacrylate): Expl
relationships between fiber formation, viscosity, molecular weight and concentration in a good solvent. *Polymer* **200**
4799–4810. [CrossRef]
41. Fridrikh, S.V.; Yu, J.H.; Brenner, M.P.; Rutledge, G.C. Controlling the fiber diameter during electrospinning. *Phys. Rev. Lett.*
90, 144502. [CrossRef]
42. Yang, G.Z.; Li, H.P.; Yang, J.H.; Wan, J.; Yu, D.G. Influence of working temperature on the formation of electrospun pol
nanofibers. *Nanoscale Res. Lett.* **2017**, *12*, 55. [CrossRef]
43. Zuo, W.; Zhu, M.; Yang, W.; Yu, H.; Chen, Y.; Zhang, Y. Experimental study on relationship between jet instability and form
of beaded fibers during electrospinning. *Polym. Eng. Sci.* **2005**, *45*, 704–709. [CrossRef]
44. Yuan, W.; Zhang, K.Q. Structural evolution of electrospun composite fibers from the blend of polyvinyl alcohol and pol
nanoparticles. *Langmuir* **2012**, *28*, 15418–15424. [CrossRef]
45. Tian, X.; Bai, H.; Zheng, Y.; Jiang, L. Bio-inspired heterostructured bead-on-string fibers that respond to environmental we
Adv. Funct. Mater. **2011**, *21*, 1398–1402. [CrossRef]
46. Fong, H.; Chun, I.; Reneker, D.H. Beaded nanofibers formed during electrospinning. *Polymer* **1999**, *40*, 4585–4592. [CrossR
47. Kadam, V.; Kyratzis, I.L.; Truong, Y.B.; Schutz, J.; Wang, L.; Padhye, R. Electrospun bilayer nanomembrane with hierarc
placement of bead-on-string and fibers for low resistance respiratory air filtration. *Sep. Purif. Technol.* **2019**, *224*, 24
[CrossRef]
48. Wang, L.; Topham, P.D.; Mykhaylyk, O.O.; Yu, H.; Ryan, A.J.; Fairclough, J.P.; Bras, W. Self-assembly-driven electrospinning
transition from fibers to intact beaded morphologies. *Macromol. Rapid Commun.* **2015**, *36*, 1437–1443. [CrossRef]
49. Megelski, S.; Stephens, J.S.; Chase, D.B.; Rabolt, J.F. Micro- and nanostructured surface morphology on electrospun po
fibers. *Macromolecules* **2002**, *35*, 8456–8466. [CrossRef]
50. Casper, C.L.; Stephens, J.S.; Tassi, N.G.; Chase, D.B.; Rabolt, J.F. Controlling surface morphology of electrospun polystyrene
Effect of humidity and molecular weight in the electrospinning process. *Macromolecules* **2004**, *37*, 573–578. [CrossRef]
51. McCann, J.T.; Marquez, M.; Xia, Y. Highly porous fibers by electrospinning into a cryogenic liquid. *J. Am. Chem. Soc.* **200**
1436–1437. [CrossRef]
52. Pai, C.L.; Boyce, M.C.; Rutledge, G.C. Morphology of porous and wrinkled fibers of polystyrene electrospun from dimet
mamide. *Macromolecules* **2009**, *42*, 2102–2114. [CrossRef]
53. Dayal, P.; Liu, J.; Kumar, S.; Kyu, T. Experimental and theoretical investigations of porous structure formation in electr
fibers. *Macromolecules* **2007**, *40*, 7689–7694. [CrossRef]
54. Chen, P.Y.; Tung, S.H. One-step electrospinning to produce nonsolvent-induced macroporous fibers with ultrahigh oil adso
capability. *Macromolecules* **2017**, *50*, 2528–2534. [CrossRef]
55. Qi, Z.; Yu, H.; Chen, Y.; Zhu, M. Highly porous fibers prepared by electrospinning a ternary system of nonsolvent/solvent/p
lactic acid). *Mater. Lett.* **2009**, *63*, 415–418. [CrossRef]

Rezabeigi, E.; Wood-Adams, P.M.; Demarquette, N.R. Complex morphology formation in electrospinning of binary and ternary poly(lactic acid) solutions. *Macromolecules* **2018**, *51*, 4094–4107. [CrossRef]

Lubasova, D.; Martinova, L. Controlled morphology of porous polyvinyl butyral nanofibers. *J. Nanomater.* **2011**, *2011*, 292516. [CrossRef]

Shu, D.; Xi, P.; Li, S.; Li, C.; Wang, X.; Cheng, B. Morphologies and properties of PET nano porous luminescence fiber: Oil absorption and fluorescence-indicating functions. *ACS Appl. Mater. Interfaces* **2018**, *10*, 2828–2836. [CrossRef]

Katsogiannis, K.A.G.; Vladisavljevic, G.T.; Georgiadou, S. Porous electrospun polycaprolactone (PCL) fibres by phase separation. *Eur. Polym. J.* **2015**, *69*, 284–295. [CrossRef]

Li, D.; Wu, T.; He, N.; Wang, J.; Chen, W.; He, L.; Huang, C.; El-Hamshary, H.A.; Al-Deyab, S.S.; Ke, Q.; et al. Three-dimensional polycaprolactone scaffold via needleless electrospinning promotes cell proliferation and infiltration. *Colloids Surf. B-Biointerfaces* **2014**, *121*, 432–443. [CrossRef]

Seo, Y.A.; Pant, H.R.; Nirmala, R.; Lee, J.H.; Song, K.G.; Kim, H.Y. Fabrication of highly porous poly (ε-caprolactone) microfibers via electrospinning. *J. Porous Mater.* **2011**, *19*, 217–223. [CrossRef]

Wang, H.; Yang, X.; Wu, Q.; Zhang, Q.; Chen, H.; Jing, H.; Wang, J.; Mi, S.; Rogach, A.L.; Niu, C. Encapsulating silica/antimony into porous electrospun carbon nanofibers with robust structure stability for high-efficiency lithium storage. *ACS Nano* **2018**, *12*, 3406–3416. [CrossRef] [PubMed]

Liu, J.; Xiong, Z.; Wang, S.; Cai, W.; Yang, J.; Zhang, H. Structure and electrochemistry comparison of electrospun porous carbon nanofibers for capacitive deionization. *Electrochim. Acta* **2016**, *210*, 171–180. [CrossRef]

Gupta, A.; Saquing, C.D.; Afshari, M.; Tonelli, A.E.; Khan, S.A.; Kotek, R. Porous nylon-6 fibers via a novel salt-induced electrospinning method. *Macromolecules* **2009**, *42*, 709–715. [CrossRef]

Ishita, I.; Singhal, R. Porous multi-channel carbon nanofiber electrodes using discarded polystyrene foam as sacrificial material for high-performance supercapacitors. *J. Appl. Electrochem.* **2020**, *50*, 809–820. [CrossRef]

Tran, C.; Kalra, V. Fabrication of porous carbon nanofibers with adjustable pore sizes as electrodes for supercapacitors. *J. Power Sources* **2013**, *235*, 289–296. [CrossRef]

Munoz-Bonilla, A.; Fernandez-Garcia, M.; Rodriguez-Hernandez, J. Towards hierarchically ordered functional porous polymeric surfaces prepared by the breath figures approach. *Prog. Polym. Sci.* **2014**, *39*, 510–554. [CrossRef]

Niu, C.; Meng, J.; Wang, X.; Han, C.; Yan, M.; Zhao, K.; Xu, X.; Ren, W.; Zhao, Y.; Xu, L.; et al. General synthesis of complex nanotubes by gradient electrospinning and controlled pyrolysis. *Nat. Commun.* **2015**, *6*, 7402. [CrossRef] [PubMed]

Tian, Q.; Wu, N.; Wang, B.; Wang, Y. Fabrication of hollow SiC ultrafine fibers by single-nozzle electrospinning for high-temperature thermal insulation application. *Mater. Lett.* **2019**, *239*, 109–112. [CrossRef]

Tang, Y.; Liu, Z.; Zhao, K. Fabrication of hollow and porous polystyrene fibrous membranes by electrospinning combined with freeze-drying for oil removal from water. *J. Appl. Polym. Sci.* **2019**, *136*, 47262. [CrossRef]

Li, Z.; Zhang, J.T.; Chen, Y.M.; Li, J.; Lou, X.W. Pie-like electrode design for high-energy density lithium-sulfur batteries. *Nat. Commun.* **2015**, *6*, 8850. [CrossRef]

Zhang, H.; Liu, Y.; Cui, K.; Zhao, Q.; Huang, J.; Mao, S.; Jiang, T.; Ma, Z. Electrospun ribbon-like microfiber films of a novel guanidine-based ABA triblock copolymer: Fabrication, antibacterial activity, and cytotoxicity. *Macromol. Chem. Phys.* **2019**, *220*, 1900138. [CrossRef]

Celebioglu, A.; Uyar, T. Electrospun porous cellulose acetate fibers from volatile solvent mixture. *Mater. Lett.* **2011**, *65*, 2291–2294. [CrossRef]

Ma, C.; Cao, E.; Li, J.; Fan, Q.; Wu, L.; Song, Y.; Shi, J. Synthesis of mesoporous ribbon-shaped graphitic carbon nanofibers with superior performance as efficient supercapacitor electrodes. *Electrochim. Acta* **2018**, *292*, 364–373. [CrossRef]

Afshari, E.; Mazinani, S.; Ranaei-Siadat, S.O.; Ghomi, H. Surface modification of polyvinyl alcohol/malonic acid nanofibers by gaseous dielectric barrier discharge plasma for glucose oxidase immobilization. *Appl. Surf. Sci.* **2016**, *385*, 349–355. [CrossRef]

Guo, Q.; Liu, L.; Wu, T.; Wang, Q.; Wang, H.; Liang, J.; Chen, S. Flexible and conductive titanium carbide-carbon nanofibers for high-performance glucose biosensing. *Electrochim. Acta* **2018**, *281*, 517–524. [CrossRef]

Lu, N.; Shao, C.; Li, X.; Shen, T.; Zhang, M.; Miao, F.; Zhang, P.; Zhang, X.; Wang, K.; Zhang, Y.; et al. CuO/Cu_2O nanofibers as electrode materials for non-enzymatic glucose sensors with improved sensitivity. *RSC Adv.* **2014**, *4*, 31056. [CrossRef]

Cruz-Navarro, J.A.; Hernandez-Garcia, F.; Romero, G.A.A. Novel applications of metal-organic frameworks (MOFs) as redox-active materials for elaboration of carbon-based electrodes with electroanalytical uses. *Coord. Chem. Rev.* **2020**, *412*, 213263. [CrossRef]

Shi, S.H.; Zhou, D.C.; Jiang, Y.H.; Cheng, F.C.; Sun, J.P.; Guo, Q.Q.; Luo, Y.T.; Chen, Y.G.; Liu, W. Lightweight Zn-philic 3D-Cu scaffold for customizable zinc ion batteries. *Res. Adv. Funct. Mater.* **2024**, 2312664. [CrossRef]

Hou, S.; Lu, N.; Zhu, Y.; Zhang, J.; Zhang, X.; Yan, Y.; Zhang, P.; Zhang, Z. Photoinduced phase-transition on CuO electrospun nanofibers over the TiO_2 photosensitizer for enhancing non-enzymatic glucose-sensing performance. *J. Alloys Comp.* **2022**, *900*, 163409. [CrossRef]

Chen, T.; Liu, D.; Lu, W.; Wang, K.; Du, G.; Asiri, A.M.; Sun, X. Three-dimensional Ni_2P nanoarray: An efficient catalyst electrode for sensitive and selective nonenzymatic glucose sensing with high specificity. *Anal. Chem.* **2016**, *88*, 7885–7889. [CrossRef] [PubMed]

82. Shi, H.Y.; Wu, Y.; Wang, W.; Song, W.B.; Liu, T.M. Dopant-stimulated CuO nanofibers for electro-oxidation and determinatio glucose. *Chem. Res. Chin. Univ.* **2013**, *29*, 861–867. [CrossRef]
83. Saravanan, J.; Vignesh, A.; Shah, S.S.; Aziz, M.A.; Pannipara, M.; Al-Sehemi, A.G.; Phang, S.M.; Ng, E.L.; Abdul, B.A.A.; Kun G.G. Binder-less and free-standing Co-Fe metal nanoparticles-decorated PVdF-HFP nanofiber membrane as an electrochemi probe for enzyme-less glucose sensors. *Res. Chem. Intermed.* **2022**, *48*, 101–116. [CrossRef]
84. Luo, G.X.; Liu, J.L.; Xie, J.D.; Jing, J.Q.; Li, M.; Zhao, L.B.; Li, Z.K.; Yang, P.; Jiang, Z.D. A highly electrocatalytic, stretcha and breathable enzyme-free electrochemical patch based on electrospun fibers decorated with platinum nano pine needles continuous glucose sensing in neutral conditions. *Dalton Trans.* **2023**, *52*, 12988–12998. [CrossRef] [PubMed]
85. Li, M.; Huan, K.; Deng, D.; Yan, X.; Li, Y.; Luo, L. Coaxial electrospinning synthesis of size-tunable CuO/NiO hollow heterost tured nanofibers: Towards detection of glucose level in human serum. *Colloids Surf. B Biointerfaces* **2023**, *22*, 13047. [CrossR [PubMed]
86. Li, M.; Dong, J.; Deng, D.; Ouyang, X.; Yan, X.; Liu, S.; Luo, L. Mn_3O_4/NiO nanoparticles decorated on carbon nanofibers as enzyme-free electrochemical sensor for glucose detection. *Biosensors* **2023**, *13*, 264. [CrossRef] [PubMed]
87. Kim, S.; Yoon, J.; Tae, H.; Muthurasu, A. Electrospun manganese-based metal-organic frameworks for MnOx nanostructu embedded in carbon nanofibers as a high-performance nonenzymatic glucose sensor. *ACS Omega* **2023**, *8*, 42689–42698. [Cross
88. Brito, T.; Butto-Miranda, N.; Neira-Carrillo, A.; Bollo, S.; Ruiz-Leon, D. Synergistic effect of composite nickel phosph nanoparticles and carbon fiber on the enhancement of salivary enzyme-free glucose sensing. *Biosensors* **2023**, *13*, 49. [Cross [PubMed]
89. Zhang, Y.; Luo, L.; Zhang, Z.; Ding, Y.; Liu, S.; Deng, D.; Zhao, H.; Chen, Y. Synthesis of $MnCo_2O_4$ nanofibers by electrospinn and calcination: Application for a highly sensitive non-enzymatic glucose sensor. *J. Mater. Chem. B* **2014**, *2*, 529–535. [Cross
90. Huan, K.; Li, Y.; Deng, D.; Wang, H.; Wang, D.; Li, M.; Luo, L. Composite-controlled electrospinning of CuSn bimeta nanoparticles/carbon nanofibers for electrochemical glucose sensor. *Appl. Surf. Sci.* **2022**, *573*, 151528. [CrossRef]
91. Cao, F.; Zhou, Y.; Wu, J.; Li, W.; Zhang, C.; Ni, G.; Gui, P.; Song, C. Electrospinning one-dimensional surface-phosphori CuCo/C nanofibers for enzyme-free glucose sensing. *New J. Chem.* **2022**, *46*, 11531. [CrossRef]
92. Qu, L.L.; Liu, Y.Y.; He, S.H.; Chen, J.Q.; Liang, Y.; Li, H.T. Highly selective and sensitive surface enhanced Raman scatte nanosensors for detection of hydrogen peroxide in living cells. *Biosens. Bioelectron.* **2016**, *77*, 292–298. [CrossRef] [PubMed]
93. Ju, J.; Chen, W. In situ growth of surfactant-free gold nanoparticles on nitrogen-doped graphene quantum dots for electrochem detection of hydrogen peroxide in biological environments. *Anal. Chem.* **2015**, *87*, 1903–1910. [CrossRef] [PubMed]
94. Zhang, Y.; Wang, Z.; Ji, Y.; Liu, S.; Zhang, T. Synthesis of Ag nanoparticle–carbon nanotube–reduced graphene oxide hybrids highly sensitive non-enzymatic hydrogen peroxide detection. *RSC Adv.* **2015**, *5*, 39037–39041. [CrossRef]
95. Celebi, M.S.; Kara, S.K. Non-enzymatic hydrogen peroxide sensor based on palladium-decorated poly(thionine) modified gl carbon electrode. *J. Chin. Chem. Soc.* **2024**, *71*, 482–492. [CrossRef]
96. Liu, C.; Hu, J.; Yang, W.; Shi, J.; Chen, Y.; Fan, X.; Gao, W.; Cheng, L.; Luo, Q.Y.; Zhang, M. Carbon dot enhanced peroxidase- activity of platinum nanozymes. *Nanoscale* **2024**, *16*, 4637–4646. [CrossRef] [PubMed]
97. Huang, J.; Wang, D.; Hou, H.; You, T. Electrospun palladium nanoparticle-loaded carbon nanofibers and their electrocata activities towards hydrogen peroxide and NADH. *Adv. Funct. Mater.* **2008**, *18*, 441–448. [CrossRef]
98. Zhang, P.; Zhao, X.; Zhang, X.; Lai, Y.; Wang, X.; Li, J.; Wei, G.; Su, Z. Electrospun doping of carbon nanotubes and plati nanoparticles into the beta-phase polyvinylidene difluoride nanofibrous membrane for biosensor and catalysis applications. *Appl. Mater. Interfaces* **2014**, *6*, 7563–7571. [CrossRef] [PubMed]
99. Guan, H.; Zhao, Y.; Zhang, J.; Liu, Y.; Yuan, S.; Zhang, B. Uniformly dispersed PtNi alloy nanoparticles in porous N-do carbon nanofibers with high selectivity and stability for hydrogen peroxide detection. *Sens. Actuators B Chem.* **2018**, *261*, 354- [CrossRef]
100. Ding, Y.; Wang, Y.; Li, B.; Lei, Y. Electrospun hemoglobin microbelts based biosensor for sensitive detection of hydrogen pero and nitrite. *Biosens. Bioelectron.* **2010**, *25*, 2009–2015. [CrossRef]
101. Qu, X.; Zhao, S.; Gao, P.; Qian, X.; Lu, S.; Duan, F.; Zhu, H.; Du, M. Ultrasensitive hydrogen peroxide electrochemical se based on dual-phase perovskite oxide tubular nanofiber. *New J. Chem.* **2023**, *47*, 1540–1547. [CrossRef]
102. Bi, C.; Jin, K.; Yan, J.; Qin, Y.; Hong, F.; Huang, W.; Liu, Y. Nanofiber-based stretchable electrodes for oriented culture mechanotransduction monitoring of smooth muscle cells. *ACS Sensors* **2023**, *8*, 3248–3256. [CrossRef] [PubMed]
103. Li, Y.; Yin, P.; Zhang, Y.; Zhang, R. Synthesis of honeycomb Ag@CuO nanoparticles and their application as a highly sensitive electrocatalytically active hydrogen peroxide sensor material. *Anal. Methods* **2022**, *14*, 4842–4850. [CrossRef] [PubMed]
104. Zhao, X.; Zhang, Z.; Song, N.; Shi, J.; Yang, N.; Nie, G.; Wang, C. Vanadium/cobalt oxides-anchored flexible carbon nanof with tunable magnetism as recoverable peroxidase-like catalysts. *Mater. Today Chem.* **2021**, *22*, 100568. [CrossRef]
105. Riaz, M.; Yuan, Z.; Mahmood, A.; Liu, F.; Sui, X.; Chen, J.; Huang, Q.; Liao, X.; Wei, L.; Chen, Y. Hierarchically porous ca nanofibers embedded with cobalt nanoparticles for efficient H_2O_2 detection on multiple sensor platforms. *Sens. Actuators B-* **2020**, *319*, 128243. [CrossRef]
106. Shekh, M.I.; Amirian, J.; Du, B.; Kumar, A.; Sharma, G.; Stadler, F.J.; Song, J. Electrospun ferric ceria nanofibers blended MWCNTs for high-performance electrochemical detection of uric acid. *Ceram. Int.* **2020**, *46*, 9050–9064. [CrossRef]

. Mercante, L.A.; Pavinatto, A.; Iwaki, L.E.; Scagion, V.P.; Zucolotto, V.; Oliveira, O.N., Jr.; Mattoso, L.H.C.; Correa, D.S. Electrospun polyamide 6/poly(allylamine hydrochloride) nanofibers functionalized with carbon nanotubes for electrochemical detection of dopamine. *ACS Appl. Mater. Interfaces* **2015**, *7*, 4784–4790. [CrossRef]
. Liu, Y.; Huang, J.; Hou, H.; You, T. Simultaneous determination of dopamine, ascorbic acid and uric acid with electrospun carbon nanofibers modified electrode. *Electrochem. Commun.* **2008**, *10*, 1431–1434. [CrossRef]
Tang, X.; Liu, Y.; Hou, H.; You, T. Electrochemical determination of L-tryptophan, L-tyrosine and L-cysteine using electrospun carbon nanofibers modified electrode. *Talanta* **2010**, *80*, 2182–2186. [CrossRef]
Huang, J.; Liu, Y.; Hou, H.; You, T. Simultaneous electrochemical determination of dopamine, uric acid and ascorbic acid using palladium nanoparticle-loaded carbon nanofibers modified electrode. *Biosens. Bioelectron.* **2008**, *24*, 632–637. [CrossRef]
Huang, Y.; Miao, Y.E.; Ji, S.; Tjiu, W.W.; Liu, T. Electrospun carbon nanofibers decorated with Ag-Pt bimetallic nanoparticles for selective detection of dopamine. *ACS Appl. Mater. Interfaces* **2014**, *6*, 12449–12456. [CrossRef] [PubMed]
Ratlam, C.; Phanichphant, S.; Sriwichai, S. Development of dopamine biosensor based on polyaniline/carbon quantum dots composite. *J. Polym. Res.* **2020**, *27*, 183. [CrossRef]
Samie, H.; Arvand, M. RuO_2 nanowires on electrospun CeO_2-Au nanofibers/functionalized carbon nanotubes/graphite oxide nanocomposite modified screen-printed carbon electrode for simultaneous determination of serotonin, dopamine and ascorbic acid. *J. Alloys Comp.* **2019**, *782*, 824–836. [CrossRef]
Yin, Z.; Ji, Z.; Bloom, B.; Jayapalan, A.; Liu, M.; Zeng, X.; Waldeck, D.; Wei, J. Manipulating cobalt oxide on N-doped aligned electrospun carbon nanofibers towards instant electrochemical detection of dopamine secreted by living cells. *Appl. Surf. Sci.* **2022**, *577*, 151912. [CrossRef]
Veeralingam, S.; Badhulika, S. Surface functionalized β-Bi_2O_3 nanofibers based flexible, field-effect transistor-biosensor (BioFET) for rapid, label-free detection of serotonin in biological fluids. *Sens. Actuators B Chem.* **2020**, *321*, 128540. [CrossRef]
Mondal, K.; Ali, M.A.; Agrawal, V.V.; Malhotra, B.D.; Sharma, A. Highly sensitive biofunctionalized mesoporous electrospun TiO_2 nanofiber based interface for biosensing. *ACS Appl. Mater. Interfaces* **2014**, *6*, 2516–2527. [CrossRef]
Atilgan, H.; Unal, B.; Yalcinkaya, E.E.; Evren, G.; Atik, G.; Kirbay, F.O.; Kilic, N.M.; Odaci, D. Development of an enzymatic biosensor using glutamate oxidase on organic-inorganic-structured, electrospun nanofiber-modified electrodes for monosodium glutamate detection. *Biosensors* **2023**, *13*, 430. [CrossRef]
Macwan, I.; Aphale, A.; Bhagvath, P.; Prasad, S.; Patra, P. Detection of cardiovascular CRP protein biomarker using a novel nanofibrous substrate. *Biosensors* **2020**, *10*, 72. [CrossRef]
Yin, Y.; Zhao, X.; Xiong, J. Modeling analysis of silk fibroin/poly(ε-caprolactone) nanofibrous membrane under uniaxial tension. *Nanomaterials* **2019**, *9*, 1149. [CrossRef]
Schubert, D.W. Revealing novel power laws and quantization in electrospinning considering jet splitting—Toward predicting fiber diameter and its distribution. *Macromol. Theory Simul.* **2019**, *28*, 1900006. [CrossRef]
Fashandi, H.; Karimi, M. Comparative studies on the solvent quality and atmosphere humidity for electrospinning of nanoporous polyetherimide fibers. *Ind. Eng. Chem. Res.* **2013**, *53*, 235–245. [CrossRef]
Maynard, A.D.; Aitken, R.J.; Butz, T.; Colvin, V.; Donaldson, K.; Oberdorster, G.; Philbert, M.A.; Ryan, J.; Seaton, A.; Stone, V.; et al. Safe handling of nanotechnology. *Nature* **2006**, *444*, 267–269. [CrossRef] [PubMed]
Yurova, N.S.; Danchuk, A.; Mobarez, S.N.; Wongkaew, N.; Rusanova, T.; Baeumner, A.J.; Duerkop, A. Functional electrospun nanofibers for multimodal sensitive detection of biogenic amines in food via a simple dipstick assay. *Anal. Bioanal. Chem.* **2018**, *410*, 1111–1121. [CrossRef] [PubMed]
Cho, H.J.; Kim, Y.H.; Park, S.; Kim, I.D. Design of hollow nanofibrous structures using electrospinning: An aspect of chemical sensor applications. *ChemNanoMat* **2020**, *6*, 1014–1027. [CrossRef]
Dong, C.; Fu, Y.; Zang, W.; He, H.; Xing, L.; Xue, X. Self-powering/self-cleaning electronic-skin basing on PVDF/TiO_2 nanofibers for actively detecting body motion and degrading organic pollutants. *Appl. Surf. Sci.* **2017**, *416*, 424–431. [CrossRef]
Wang, X.; Song, W.Z.; You, M.H.; Zhang, J.; Yu, M.; Fan, Z.; Ramakrishna, S.; Long, Y. Bionic single-electrode electronic skin unit based on piezoelectric nanogenerator. *ACS Nano* **2018**, *12*, 8588–8596. [CrossRef] [PubMed]
Liu, S.; Li, L.; Liu, R.; Liu, J.; Zhao, X.; Liao, X. Ultra-flexibility, robust waterproof and breathability of polyvinylidene fluoride membrane blended silver nitrate for electronic skin substrate. *J. Phys. D Appl. Phys.* **2020**, *53*, 195403. [CrossRef]
Teodoro, K.B.R.; Shimizu, F.M.; Scagion, V.P.; Correa, D.S. Ternary nanocomposites based on cellulose nanowhiskers, silver nanoparticles and electrospun nanofibers: Use in an electronic tongue for heavy metal detection. *Sens. Actuators B Chem.* **2019**, *290*, 387–395. [CrossRef]
Oliveira, J.E.; Scagion, V.P.; Grassi, V.; Correa, D.S.; Mattoso, L H.C. Modification of electrospun nylon nanofibers using layer-by-layer films for application in flow injection electronic tongue: Detection of paraoxon pesticide in corn crop. *Sens. Actuators B Chem.* **2012**, *171–172*, 249–255. [CrossRef]
Atas, H.B.; Kenar, A.; Tastekin, M. An electronic tongue for simultaneous determination of Ca^{2+}, Mg^{2+}, K^+ and NH_4^+ in water samples by multivariate calibration methods. *Talanta* **2020**, *217*, 121110. [CrossRef]

MDPI AG
Grosspeteranlage 5
4052 Basel
Switzerland
Tel.: +41 61 683 77 34

Molecules Editorial Office
E-mail: molecules@mdpi.com
www.mdpi.com/journal/molecules

www.ingramcontent.com/pod-product-compliance
Lightning Source LLC
Chambersburg PA
CBHW062037220926
51850CB00010B/2579
9783725875870